T0371992

Energy Independence:
The Individual Pursuit of
Energy Freedom

RIVER PUBLISHERS SERIES IN ENERGY SUSTAINABILITY AND EFFICIENCY

Series Editors

PEDRAM ASEF
Lecturer (Asst. Prof.) in Automotive Engineering,
University of Hertfordshire,
UK

The "River Publishers Series in Sustainability and Efficiency" is a series of comprehensive academic and professional books which focus on theory and applications in sustainable and efficient energy solutions. The books serve as a multi-disciplinary resource linking sustainable energy and society, fulfilling the rapidly growing worldwide interest in energy solutions. All fields of possible sustainable energy solutions and applications are addressed, not only from a technical point of view, but also from economic, social, political, and financial aspects. Books published in the series include research monographs, edited volumes, handbooks and textbooks. They provide professionals, researchers, educators, and advanced students in the field with an invaluable insight into the latest research and developments.

Topics covered in the series include, but are not limited to:

- Sustainable energy development and management;
- Alternate and renewable energies;
- Energy conservation;
- Energy efficiency;
- Carbon reduction;
- Environment.

For a list of other books in this series, visit www.riverpublishers.com

Energy Independence: The Individual Pursuit of Energy Freedom

Alden M. Hathaway II

Tripp Hathaway

LONDON AND NEW YORK

We originally conceived the book title to be Energy Freedom. However in light of the events in Europe in early 2022, we recognized the need for countries to secure their energy independence from hostile nations. This book provides the roadmap for families to achieve that Energy Independence one household at a time.

Published 2022 by River Publishers
River Publishers
Alsbjergvej 10, 9260 Gistrup, Denmark
www.riverpublishers.com

Distributed exclusively by Routledge
4 Park Square, Milton Park, Abingdon, Oxon OX14 4RN
605 Third Avenue, New York, NY 10017, USA

Energy Independence: The Individual Pursuit of Energy Freedom /
Alden M Hathaway, II & Tripp Hathaway.

Routledge is an imprint of the Taylor & Francis Group, an informa business

ISBN 978-87-7022-693-6 (hardback)
ISBN 978-87-7022-767-4 (Paperback)
ISBN 978-1-003-20735-1 (ebook master)
ISBN 978-1-000-77670-6 (online)

While every effort is made to provide dependable information, the publisher,
authors, and editors cannot be held responsible for any errors or omissions.

Contents

Preface

"My grandfather, Earl Hathaway, served as President of the Firestone Tire and Rubber Company beginning in 1962. He succeeded Raymond Firestone, one of the sons of Harvey Firestone, the founder of the company. My grandfather had worked for the company since 1927, mostly in sales roles. Needless to say, he had relationships with industry leaders, such as those at the Ford Motor Company that specified his tires for their cars.

At one point, my father, an Episcopal bishop, presided over a wedding between the Firestone and Ford families. I feel a closeness to the great industry leaders of the 20th century because of my grandfather. However, I went in a different direction. I studied electrical engineering at the University of Virginia and instead of becoming a religious leader, I chose to follow in the footsteps of legends like Thomas Edison and George Westinghouse.

I mention this about my grandfather and my career to illustrate a legacy that foreshadows this book. Now, as we move through the 21st century, we bear the torch passed to us by those that accomplished so much in the 20th century. It is our turn to advance, solve the big problems and pursue the big opportunities of our day. I am honored that my work has been but a few steps in that journey."

By 1988, I'd been working in energy efficiency programs for six years, promoting Sylvania's latest lighting technologies to save energy. In those days I promoted the concept of energy savings based on payback and return on investment. In that same year, James Hansen was testifying to the US Congress that the continued use of fossil fuels would cause the Earth's temperature to increase and the ice caps to melt rising sea levels and causing mass migrations and serious changes for the world's population. Suddenly, I had a new marketing tool to use to promote energy efficiency.

I was never an ardent environmentalist, but I enjoyed nature and the outdoors, particularly the island lobstering community off Acadia National Park that was my early home. Over the first few years, after Hansen's testimony, I came to believe that we have a directive by our Creator to care for the Earth, particularly to protect it for future generations. I have seen changes in our island home over the past 50 years; tides are higher, the seas are warmer, and the propensity for road washouts due to storms seems to be increasing. I believe, our energy stewardship can be a positive response that will help move toward solving the problem of rising global temperatures.

I have been promoting energy efficiency to save money with a rate of return comparable to the top investment firms. If we can all care for the Earth, and make a positive return on investment then that is where we should put our money. For the next ten years, 1988–1998, I had the good fortune to work in the EPA's Energy Star Buildings program and learn how all the energy technologies fit together. I was also able to spend an additional year working in the solar business to learn how solar could be integrated into buildings. By 1998, I thought I knew how to build a zero-energy solar home and could do so in such a way that we would make an attractive rate of return in terms of energy cost savings on the money invested. Not only would we save energy, but we would also create it for others.

The first solar house we built was displayed on the National Mall in Washington, DC during Earth Day week in 2001 as one of the nation's first zero net energy solar homes. Dubbed the Solar Patriot, it appeared in President Bush's National Energy Policy Plan. The Solar Patriot was built in sections allowing it to be featured on the Mall and then transported to our land in Northern Virginia to be our home for seven years in the Appalachian foothills. By the time we sold the house to move to Georgia in 2007, we documented a 22% return on investment on the money we put into the energy features of the home. But, we realized that while we had been successful with our home, we had not yet fully dealt with the energy we used for our cars. At that point, we were just beginning to consider greater reduction to our carbon footprint and we had not yet fully addressed the savings potential of our vehicles.

From 2008 to 2018, we would put it all together on cars and homes to achieve zero energy in our home and significantly reduce gasoline consumption in our automobiles, such that we would reduce our energy carbon footprint by 80%, while at the same time creating an energy savings engine that would save us more than $110,000 over 20 years. As the energy savings grew each year, it eventually achieved a steady-state average of $700 per month. And that energy savings engine continues to save us an average of $700 per month, or $8,400 per year, which should double our first 20 years of savings within the next ten years.

As I am writing this, war has broken out in the Ukraine and the world is suddenly conscious of the dangers of over dependence on energy sources from autocratic nations. Where we used to just complain about rising gas prices we are now aware of potential impacts to freedom that occur when we do not have the energy flexibility we need. I often tell people who are upset about the impacts to their budgets of sudden price jumps or because of environmental and security interests to drive an electric vehicle and/or put a solar system on their house. The response is usually the same I don't drive enough or it costs too much for me to invest. But if you are frustrated with what is going on regarding our energy politics, propping up autocracies or degrading our environment, why not consider a new energy journey you can do yourself?

If I knew of a way that one could put their hard-earned dollars to work and earn a rate of return of up to 22%, or could save over $100,000 in 20 years, all while having a positive impact on our environment, wouldn't I have an obligation to share this information? I have found that creating a flow of cash from energy savings is a quite simple way to build up the cash necessary to make ongoing investments in energy efficiency. We built on the lessons of the past and achieved nearly an 80% reduction in our average monthly energy costs in less than eight years. I have already completed the task of figuring out what works best and in what order improvements should be made. I have written this book to help others plan, based on our experience with our two zero energy homes. I hope that others will take similar steps to realize their own energy savings goals.

My son, who has helped me write this book, believes that there is something uniquely American about converting your home into a zero-energy powerhouse that even powers your vehicles. Liberty is the hallmark of American life. So, it is fitting that we, as Americans, would free ourselves from the electric grid, from gasoline, from climate change, and from dependence on foreign resources. All you need to do is execute.

This is what we mean by Energy Freedom.

List of Figures

1

An Opportunity

I have a favor to ask you. What I am asking may seem strange but, bear with me.

Take a look around where you are right at this moment. What do you see? The book in your hand, the chair on which you are sitting ... look harder. Take a minute to notice some object around you or even a tangible characteristic of your environment that you would might otherwise overlook. Find something you might never consider. Find something that you might not notice that brings some value to you. Now, take some time to study it. What are its characteristics, its functions, its value? Next, make an appraisal of how that thing came to be. What had to happen to it such that became whatever it is that brings you value now.

Don't just look at the surface. Dig deep. Take the object in your mind and devolve it. Devolve it through each degree of its development, working backward, until there is nothing left but the rawest of things. Challenge yourself to go as far as you know to go. Keep going down the rabbit hole until you arrive at things that have no known origin.

What you probably realize is that even the simplest thing didn't just appear here. That object that brings you value suffered an exhaustive journey through many silent transformations and improvements, some occurring naturally and some catalyzed by man, before becoming the ignorable aspect of your current environment. Now, there's no need to put this thing on a pedestal. But this will help explain the purpose of this book and, ultimately, help you to take advantage of a great opportunity. This opportunity only exists by first solving a problem you may not even be aware of, one that we ignore in the same way you might seldom notice the thing you just finished analyzing.

But, before we get there, we should recognize a commonality in each step in the developmental process of the thing you have just studied. Can you guess which one single factor was present and contributing to each degree of the evolution of that thing? It is the same single factor present and contributing to the being of any one thing you can think of.

That factor is energy.

Think about it. There has to be some sort of input or cause present every single time something – anything – changes into something else, something better. It takes the motion of hands to force two stones together. It takes friction between the stones to produce a spark. It takes that spark to start a fire; and, it takes the fire to warm the air. Every transformation in the physical world happens because of energy. When viewed holistically, in all of its forms, energy is the most important of things in our known universe. It is the input to every output, the reason for everything. Energy is invaluable to us because we must have it to have anything.

Beyond that concept of energy, we don't need to spend any more time speaking in abstract terms. This book is practical. This book is about how we use energy every day to cook our food, to heat and cool our homes, to drive to work, to see at night, to access the internet, to make a phone call and to print this book!…

We use it as if it will never cease! It will always be there waiting for us to put it to use without any consideration that it might someday be gone.

Don't get me wrong: that's not a problem in itself. We just uncovered the fact that everything that is done in this universe is fulfilled by the incorporation of energy, so there is no way we can possibly avoid its use. And, there's no need to; energy works for our benefit. Rather, the problem comes from how we take energy for granted. We turn our TV on and it just works. The game is on, the beer is cold, and we forgot to pause it so we already missed the first minute and a half. So, we don't care how it all works. We do this time and again in our minds until we are completely oblivious to our consumption of energy.

This is the "energy as a given" mindset that we have. We think that energy is infinite and that as long as we can pay for it, we can use it at will. This is perpetuated by the fact that we never feel the consequences. So, there must not be any consequences, right? It's really hard to tell, isn't it? The slow development of the habit of taking energy for granted sedates us to this problem. But the truth is, "energy as a given" is a dangerous paradigm because it helps us ignore the cost of our *real* problems that we will discuss shortly.

Think about the rawest forms of energy. The sun shines and provides heat which warms the earth and waters, but which also provides light. The wind blows and the atmosphere churns, moving moisture from here to there. Organisms live, die, and decay, circulating organic fuel through the ecosystem. There is natural, potential energy all around us. Part of what makes mankind special is his ability to harness that potential energy for his use. We have been harnessing energy since we left the cave and we have fought for thousands upon thousands of years toward perfecting the practice ever since.

Needless to say, we are deeply invested in putting natural, potential energy to good use.

In the earlier example about building a fire, it first required someone's effort to gather the wood and to find the right spark-producing stones. Then, someone had to labor over them until a fire could be started. And, it's not just the labor of getting one fire started. Building fires is a skill that had to be learned and developed before it could be of any use to us. So, next time you are warm by a fire, think of all of the toil mankind went through in its early days to make a warm fire reality for you. Sure, the caveman has you covered on this one, but this illustrates the point that making energy useful is not free.

Suddenly, the habit of taking energy for granted seems incompatible with the cost of making it useful to us. If energy were free and infinite there would be no need to worry about waste. But, when we have to labor to transform it (or pay someone else to), waste draws a balance directly against us. Consider a warm, sunny Sunday afternoon, relaxing in your backyard with your family. It's summertime. You wouldn't even think about building a fire in your fire pit or even simply rolling out a propane heater because you don't need the heat. Even the few minutes it would take, would be a waste of your effort and a waste of your precious time with your family. We don't think this way about energy waste because we are removed from its consequences. The transformation of energy is done for us and useful energy is delivered to us without ever needing to think about it. So, we walk out of a room leaving the light on because … eh … it's only one light and we didn't want to stop to flip the switch. It shows up a month later on our utility bill just the same as it has every month. We have normalized waste, and over time our carelessness costs us a fortune.

The Opportunity

Yet therein lies our opportunity. If we can transform our waste into an engine of savings, we can invest in owning our very own energy, dependent on no one. Even better, we can be the change that our energy economy needs. But before we can talk about where we want to go, it's helpful to know where we've been.

The 20th century is known to the entire world as a century of American greatness. And innovations in energy drove that American superiority. Oil pipelines, petroleum-based fuel, the light bulb, electricity, the power grid: all of these things propelled the United States to be the top world superpower. Today, almost every household in the United States has at least one automobile; we can travel anywhere we want with more than 150,000 refueling stations across the country to keep our tanks full[1]. Electric utilities bring electricity to nearly every occupied structure in the entire nation with incredible reliability of power.

So, why are you reading this book if our energy infrastructure is so great?

It's not the 20th century anymore! Change is a part of every aspect of life, and our energy economy is no exception. There are several factors that threaten the status quo and you are the key to building our 21st-century energy grid by owning your piece of it!

As a point of clarification, when we say "energy" from this point forward we are referring to both oil and utility industries that make up our total energy economy. When you hear pundits talk about "energy" they typically are referring to oil & gas. We mean both. But because the oil & gas and utility sectors are so different from one another, we need to address them separately.

The Electric Grid Then and Now

"Great God! He's alive!" This was one of the witnesses' cries after the first attempt to execute a murderer, William Kemmler, failed gruesomely. Kemmler was finally pronounced dead after an additional agonizing attempt. The first-ever electric chair convulsed Kemmler's body; blood dripping from his coiled fists. It ignited his clothing and charred his body. For some, the electric chair was supposed to be a more humane way to carry out capital punishment. For Thomas Edison, it was something entirely different.

In the late 1800s, Thomas Edison and George Westinghouse sparred in what may be the greatest rivalry in American business history: a rivalry that would decide the foundation of more than a century of American prosperity. Edison's first light bulb worked with his direct current (DC) system for which he also had begun building generators. However, there were technical challenges with Edison's DC systems. Direct current cannot travel very far before the voltage carried in the lines drops. The longer the run of wire, the lower the voltage, eventually providing inadequate power for a light bulb just a couple of city blocks away from the generator. Committed to his inventions, Edison dismissed the advice of Nikola Tesla, whom Edison had hired to solve his technical problems. Tesla had advised Edison to switch to an alternating current (AC). Instead, Westinghouse took notice of Tesla and began bringing AC generators with compatible AC lighting to market.

---

[1] https://www.api.org/oil-and-natural-gas/consumer-information/consumer-resources/service-station-faqs#:~:text=The%20NACS%2C%20the%20association%20for,selling%20fuel%2C%20marinas%2C%20etc.

Using Tesla's technology, Westinghouse could build larger and fewer generators that could light homes farther away and quickly took market share from Edison. To fight the proliferation of Westinghouse's AC systems, Edison sued on the grounds that Westinghouse infringed on his patent for the light bulb. The colossal legal battle inevitably invited bitter publicity campaigns between the two giants. One such campaign was Edison's effort to smear Westinghouse's systems as unsafe, even deadly. Having convicts of capital crimes "Westinghoused" would have accomplished that end ... until William Kemmler's charred body vindicated the Westinghouse brand.

Westinghouse won the "Current Wars" predicating an AC grid with large, centralized generators and transmission and distribution that carries power over long distances. This centralized AC grid lent itself to greater economies of scale which made energy cheap. Cheap energy meant households could not only afford to power lights, but a household radio, television, refrigerator, electric heating, air conditioning, computers, and cell phones. Likewise, the industry could transition to electricity as a more efficient and cost-effective means of production. For homes and industry to have access to reliable power, transmission and distribution (T&D) lines had to be run all over the country. Large power plants pushed power in all directions via this T&D system.

All of the necessary infrastructure that the 20th century AC power grid required created a familiar economic entity – a natural monopoly. Not unlike the railroad and oil industries shortly before, the costs of the competition were too high to bear and natural monopoly was preferable, at least for the time being. However, weary of more economic power being heavily concentrated in yet another new industry (as had characterized the late 1800s and early 1900s) the federal and state governments began regulating utilities and even creating their own utilities. For example, the Tennessee Value Authority was created in 1933 under the FDR administration as part of the New Deal and remains today a wholly-owned government corporation.

Before addressing the effects of regulation on utility monopolies, it's helpful to first understand the economic effects of a monopoly. In perfect competition, producers are price-takers, meaning that the market decides the price and quantity of a good and that occurs where the demand and marginal cost – or supply – curves intersect (where the willingness of consumers to buy meets the willingness of producers to produce). Perfect competition provides the maximum benefit for producers and consumers. This is not so with monopolies.

The major difference between monopolies and firms in perfect competition is their marginal revenue. In perfect competition, marginal revenue equals the price at the quantity demanded. Let's say you are buying a dozen eggs on a trip to the grocery store and the price is $2. Your demand for eggs is only one dozen, if your demand were higher, you would have two dozen or more in your cart. What if there were a deal for buying another dozen? You won't turn down a BOGO offer (who says "no" to free stuff?), but you won't buy more than you need at the same price of $2. So, let's assume $1 is a low enough price to convince you to buy one more and that's the deal for the second dozen offered at the store. You paid $2 for the first and $1 for the second. The grocery store also made $2 on the first and $1 on the second. This is made possible by more than one egg producer supplying different quantities of eggs at different prices that aggregate to meet the total market demand, whether at $2 here or $1 there. Each egg producer has a different cost structure and comes to a different price to maximize their profit.

However, with a monopoly, there is only one producer, one cost structure, therefore one price. So, an increase in the quantity produced means a drop in price for all units produced, not just the marginal unit. So, the marginal revenue from the next unit sold in a monopoly is less than the price at that quantity demanded. This is the critical economic concept that establishes the other characteristics of a monopoly.

Keep in mind that this is a simple description of the natural tendency of a monopoly in economics. Today many monopoly utilities offer various rate classes for different types of customers. They may charge a homeowner the full retail rate for a kWh delivered and turn around to a large industrial user and charge them for capacity (typically in megawatts of power). They may even have volume discounts built into the rate schedule. It doesn't matter. Pricing plans offered by a single company and the economic concept of price are not the same. We are still talking about only one utility with one cost structure as the basis for its pricing.

No one can afford to produce a unit of anything that costs more than it can be sold for – not even monopolies. So, because their marginal revenue is less than what could be earned at that quantity demanded, they must produce fewer units. We know from a downward-sloping demand curve that producing fewer of something makes the price that people are willing to pay higher. So, monopolies produce less and charge more per unit than they would at equilibrium in perfect competition. Because less is produced at a higher price, society bears a loss in economic efficiency compared to equilibrium. This is called deadweight loss in economics.

Does this make monopolies bad? Not necessarily. In the case of the utilities in the 20th century, the economic losses from monopoly power were preferable to overcoming enormous barriers to entry and cluttered infrastructure if competitors did overcome them.

Does the reduced quantity produced in monopolies mean that utilities don't provide enough power? Not at all. Utilities have built the grid to produce and deliver one of the most reliable products any industry in the history of the world has ever known. And, importantly, anyone that wants access to the grid can get it. It just means that a higher quantity at a lower price would be more economically efficient.

Nevertheless, governments saw a need to manage utility monopoly power to maximize societal benefit. Utility regulation was managed by the states, which established special governing bodies, typically called public service commissions. In addition to state regulations federal legislation was passed that also addressed how utilities did business. For the general purpose of injecting some market efficiency back into the way utilities naturally operated and to keep prices down, states placed requirements on planning for generating capacity, approved rate hikes, and oversaw efforts like rural electrification. The federal government addressed broader issues. For example, the Public Utility Regulatory Policy Act of 1978 injected some competition into the generation side of the industry by encouraging third-party investment in smaller-scale generators called Qualifying Facilities from which utilities would be required to purchase power under certain conditions.

Deregulation has occurred in a handful of state beginning with Rhode Island in 1996, but the majority of markets are still regulated to this day. To be clear, deregulation applies mostly, if not exclusively, to generation. Transmission & distribution is left in the hands of a single grid operator. The generation side of the industry was ready for open competition in 1978. If third parties were willing to invest in generators for power sales to utilities, then the barriers-to-entry characteristic of natural monopolies had reduced. Two interesting questions arose. First, why did it take almost two decades for the first state to deregulate? Second, why more than four decades later, are most markets still regulated?

We have long outgrown the need for regulation. Bad actors like Enron made many think twice about deregulation. However, the real reason is that deconstructing the web of government involvement in an enormous and infrastructure-intensive industry takes a lot of time.

You don't have to wait.

Oil & Gas Then and Now

"Thar' she blows!" A maniacal and vengeful Captain Ahab yells upon sighting the sperm whale that took his leg.

Okay, maybe there isn't a true story about the rise of petroleum that is quite as riveting as the first electric chair story. That is why we turned to fiction in Moby Dick. But, the timing of Moby Dick's story being published in 1851 offers some insight as to why it is so pertinent to the petroleum industry. The centuries-old whaling industry was seeing strains of scarcity in the 19th century. Whalers could no longer find whales near shore in small vessels. Larger vessels and more crew were needed. Did Captain Ahab take his crew and ship back out to sea for just one whale because of his mad desire for revenge or was such an expense typical for a dwindling return of whale by-products at that time?

Whales were harvested for their baleen which are their tooth-like krill filter. They also were popular for their bones, and fat. Their fat in particular was cooked down into whale oil. Whale oil was primarily used for lighting and as a lubricant. But, as whales grew scarce and the cost of whale oil spiked in the 1800s, other alternatives came to market. Lard oil from livestock and alcohol were common substitutes for lamp fuel, for example. Precipitously, the first petroleum well successfully drilled in 1859 in Titusville, PA gave way to the proliferation of a new kind of oil that was destined to replace whale oil. Even though this new oil seemed to be a replacement, there were problems associated with the logistics of producing the petroleum oil. It became clear quickly that the problems associated with replacing whale oil were not yet solved. The logistics of petroleum were not at all like the logistics of whale oil. For example, once the whale's blubber had been harvested, there was a fixed quantity to cook down into oil and then transport to market. The oil could be produced on the ship even before returning to port. In the case of Petroleum, one of the problems is that it is generally free-flowing from the ground. It is not contained within a captured animal. This product must be captured when the pump is running. It simply has to go somewhere.

The challenge of moving oil eventually found pipelines as a solution. Samuel Van Syckel built the first pipeline, just long enough to get oil from an oil well to a nearby train station. Pipelines remained short and used only for local connections until John D. Rockefeller. Rockefeller, America's first billionaire and founder of Standard Oil, is credited with making the pipeline the primary means of transporting oil from wells all the way to refineries. The new oil infrastructure that Rockefeller helped to create primed the industry to be able to serve new demands for petroleum. It was first used to power gasoline and diesel engines for the automobile and later for just about every other mode of transportation as well as for electric power generation.

As we consider oil and gasoline, our focus is solely on transportation and mainly on personal automobiles. Electricity would soon replace oil as a means of providing lighting. The use of oil in power generation has been merely one of many fuels for the utility industry. It is, therefore, further away from our concern as consumers since we focus on the plug in the wall without concern as to how the power was produced or brought there. The first gasoline-powered automobile was invented around 1875 by Siegfried Marcus. The exact date and some details are not certain as Nazi Germany attempted to destroy Marcus' legacy since he was a Jewish man. Instead, Carl Benz is often credited as the inventor of the first gasoline-powered car in 1886. In the following years, many advancements were made that added to the automobile. But Henry Ford made the automobile commonplace. According to Ford Motor Company, in their own chronicles, "by the early 1920s "[m]ore than 15,000,000 Model T's were built and sold." Today, bts.gov in their last data set in 2020 275.9 million highway vehicles were registered in the US. This only includes personal vehicles, motorcycles, lightweight utility vehicles, and buses.

From Ford's Model T to today's vehicles, the range of transportation that uses internal combustion engines or other types of engines that burn petroleum fuels far out-paces highway vehicles alone. Think of all of the possible ways to move people and goods around the Earth. Highway transportation is overwhelmingly petroleum-fueled, which includes diesel and liquefied natural gas (a very small percentage is Liquid Natural Gas). Commercial aviation is entirely petroleum-fueled. Almost all of our railroads are occupied by mostly diesel-powered engines with only a small number of electric passenger trains making up the total rail traffic. Originally powered by renewable energy (sail), ocean-fairing freight vessels are overwhelmingly powered by marine heavy fuel oil (HFO), although some hydrogen fuel is being used and the industry is considering nuclear propulsion. The bottom line: not only do we rely on petroleum for our personal transportation, we rely on petroleum for just about everything.

Politically, our significant dependence on petroleum has elicited strategies from bolstering oil production domestically to ensuring supply from abroad. In the last couple of decades, we've even seen growing demand for alternative fuels that reduce or eliminate our dependency on fossil fuels entirely. If we think about it at all (and we should) our dilemma is not that different! We must have fuel for almost every aspect of our life.

The concept of demand elasticity applies here, so it will be helpful to compare a gallon of gas to another commodity … maybe a gallon of milk. For a decade before the recent oil glut beginning late 2015 through most of 2020, we experienced a couple of unpleasant spikes in gas prices. The national average peaked as high as $4.50 per gallon and some states saw prices as high as $6.00 per gallon. During those times some people who weren't all that concerned could be heard challenging the rest of us. They would often say something like, "Well, you don't have a problem paying $3.50 for a gallon of milk, do you?" What a silly question! Milk and gasoline are worlds apart from each other. The utility we get from milk is totally different from the utility we get from gasoline. So, any sort of comparison on price between the two is pointless. However, their stark differences in utility do offer a poignant comparison in demand elasticity, as shown in Figures 1 and 2, below:

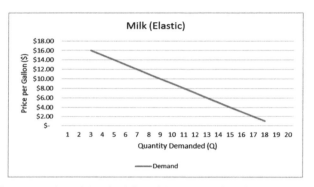

Figure 1.   Demonstration of the elasticity of the demand for milk vs the price per gallon.

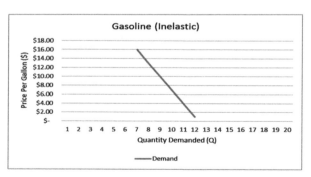

Figure 2.   Demonstration of the relative inelasticity of the demand for gasoline vs the price per gallon.

To set the scenario, let's assume that a gallon of milk sells for $3.50 and a gallon of gas sells for $2.50. Because we consume more gallons of gasoline than milk we will compare how many gallons of milk we will buy in the next six months to how many gallons of gasoline we will buy on our next fill-up. This type of comparison wil ensure the quantities demanded will be within the same range. If the price of milk goes up $10 per gallon to a price of $13.50, most of us will decide we won't buy milk unless we are compelled to do so. We will often decide to buy a substitute like soy milk or juice. We may even decide to avoid buying milk at all. If the price of gas goes up $10 per gallon to a price of $12.50, we may alter our driving habits and the amount of driving we do as much as possible in order to reduce the amount of gas we need to buy. But, we are still going to drive to the places we need to go. Going to work each day is one of those places we need to go. In the case of work, we have to drive there in order to pay for the rediculous $12.50 per gallon gasoline.

This is the concept of demand elasticity. The demand curve for a given commodity is considered elastic if the quantity demanded responds quickly to a slight movement in price. The opposite is true for an inelastic demand curve, like that of gasoline, or petroleum as a whole. In the case of inelastic demand, the quantity demanded responds to a relatively lesser degree, even to large movements in price. Granted, the demand curve is more of a true curve in reality. In other words, with gas at $12.50 per gallon, our demand will become more elastic as we desperately find new ways to avoid buying it. Either way, the concept is true. We cannot decide to not buy gasoline in the same way we can decide to not buy milk. In the short run, we have little choice but to continue to fill up our tanks when we need to regardless of the price. This is what dependence looks like.

Dependence is weakness. But you can put yourself in a position of strength.

2

Cost of the Status Quo

We have established that there are two layers to our opportunity. On the macro scale, the status quo is begging to erupt into a new and better energy economy that you will help create. On the micro-scale, investing in your own energy will not only make you less dependent on others for critical resources, but it will also be a source of savings with returns that rival your retirement investing. Where does this savings come from? Savings are achieved by reducing costs. We must first reduce the cost of our own waste and eventually replace the energy we buy with the energy we produce more cheaply. But it gets bigger than that. If more energy users do what you are about to do, our grid can be transformed into a free energy marketplace and we can finally end our dependence on oil.

It starts with us and the "energy as a given" mindset from Chapter 1. This mindset means that our demand for energy is maximized, adding our wastefulness on top of what we would reasonably consume. Especially in the first world, we have the luxury of not having to worry about waste. We no longer have time to consider how that nearby cell tower connects our cell phone to the world with a 5G signal. We are late for a meeting and have to make a call! We don't notice our wastefulness because we take energy for granted, and we just think that the expense we pay is normal. And it is normal because we have made it that way.

Just because we take energy for granted, doesn't mean converting raw potential into something useful is any easier. Energy suppliers still must do the important job of delivering it, which they will do for whatever price required. Our lack of concern for where energy comes from combined with the inelasticity of demand for oil and the monopoly power of many utilities elevates the cost of our waste even more. We complain about the power bill or the price at the gas pump. Do we do anything about it? For the most part, no.

So, we have almost completely handed over the concern of our energy to energy suppliers. Cost should be no surprise to us.

Cost of the Status Quo: Electricity

The utility industry has built an impressive supply chain to deliver pure electricity right to our electrical panels. Because of the inelasticity of our demand for energy, this network must be consistently reliable so that we are never left in the dark. Energy providers have woven redundancy, resiliency, and reserves into our energy supply chain because we simply cannot live without it. What would happen if our access to energy suddenly disappeared? The slightest interruption in our energy supply disrupts our very lives.

The electric grid is not perfect, though, and outages do occur. When a strong storm comes through and takes down power lines, or a critter crawls into a transformer, or utility equipment malfunctions, the power can simply go out. An all-electric kitchen will shut down, there's no TV or internet or light, and the HVAC system will not run. For example, that last one in the South in the summertime or the Northern Midwest in the dead of winter created havoc and cost lives. Outages can last for days or even weeks in some extreme cases if there is a major event like a hurricane, blizzard, or wildfire. The good news is our electric utilities and cooperatives are exceptional at mending the damage and getting the power back on very quickly so that we can resume our lives.

"Is this mayonnaise still good?"

It's true that on *average* outage frequency and duration appear negligible. The utility industry is measured by IEEE (Institute of Electrical and Electronics Engineers) standards of reliability. These standards include average outages and average outage duration per customer. According to the Energy Information Administration,[2] in 2019 (2020 not used due to COVID-related abnormalities) Americans experienced 1.65 outages with an average duration of 267 minutes (or four and a half hours).

An outage or two for four hours never hurt anyone right?

Averages hide the extremes. Because our grid is so centralized, if something goes wrong, multitudes of people are affected. The extremes can cause major problems and even be deadly.

That duration data includes major events. It is industry standard to report outage data associated both with major events and without them. This is a way for utilities to distinguish the outages they have the most control over from the ones they have less control over. It turns out that, looking at the data without major events, it is clear that the industry is incredibly consistent at keeping outages and outage durations very low when there isn't a major event. But, because the centralization of the grid is one of the major problems with the

---

[2] https://mentor.ieee.org/3000-stds/dcn/19/stds-19-0001-00-3006-3006-8-2018.pdf

status quo, we must also talk about the major events. Centralized generation and distribution allow major events to affect us more than they should.

In August of 2003, it was hotter than usual in First Energy's service territory, causing a few things to happen. First, loads were high as HVAC systems worked hard to beat the heat. The overhead transmission lines stretched and sagged in part because of the heat, but even more so because of the higher than usual load on the system and a nearby power plant that went offline. Coincidentally, First Energy had cut back on some of its operations and maintenance expenditures and trimmed less of the brush beneath its transmission lines. When the lines sagged near Cleveland, OH it shorted out on some of the taller brush that had not been cut. This event tripped off the transmission circuit. Compounding the issue, an alarm meant to alert grid operators of the problem, was inoperable. When the first transmission line tripped, other lines had to pick up the slack and serve more load. They could not, and they also tripped off. This created a chain reaction. Power tripped off in Michigan which was importing power from Ohio. Power tripped off across the Detroit River into Canada importing from Michigan New York went dark from across the Niagara River as far as Manhattan. Pennsylvania lost power as the outage then moved back westward. Within seconds, cities like Pittsburgh, New York City, and Toronto were dark. The only reason this blackout didn't continue across the whole country was because grid operators around the affected area quickly isolated their grids, halting the domino effect. Tens of millions were affected by the blackout.

The good news is that the outage was short-lived. No catastrophic damage was done. For most of the affected areas, generating systems that had been brought offline when the circuits tripped off simply had to be started back up and rebalanced with the load as things returned to normal. Most people had power back by the next morning. Nonetheless, the centralized nature of the grid contributed to the magnitude of this blackout that brought mayhem to the streets of New York City, 400 miles away from the point of the original fault. Imagine being trapped in an elevator or subway when the lights went out! The blackout even claimed a handful of lives. Without elevators, residents of high-rise apartments were left with the stairs contributing to the heart attack of one person. A burn victim succumbed to his injuries when air conditioning was no longer available to keep his skin grafts cool. All this because a few tree limbs 400 miles away weren't trimmed.

Just recently on August 29, 2021, Hurricane Ida made landfall south of New Orleans, Louisiana as a category 4 hurricane. Unlike Katrina 16 years earlier, Ida was not a major flooding disaster, thanks to post-Katrina improvements in the area's levees. Instead, Ida's 150 mile-per-hour winds were the threat. A spokesman from Entergy, the area's largest power provider, reported eight transmission lines were down. Entergy operates in multiple states. One of the transmission lines, according to the spokesman, was one of the most robust lines in the company which remained intact despite Katrina. However, it did not stand up to Ida and over one million people in the area would be without power, many of them for a week or more.

Ironically, the city-wide outage in New Orleans threatened plans for a major international smart energy conference that was scheduled shortly after the storm.

Transmission lines are designed for a centralized grid. Power is transformed into higher voltages which allow the use of smaller wires, cost-effectively carrying power from large plants to many different distribution circuits far away. Voltage is stepped down at each distribution circuit in order to serve the customers on that circuit. The main advantage of Westinghouse's AC over Edison's DC systems was that he could carry power from his AC dynamos further without losing power to a drop in voltage. Today, transmission lines are critical to our grid as they carry power from large power plants to customers hundreds of miles away.

We know that the centralization of our grid is now its very weakness. But, what do utilities do about it? Since Katrina, smart energy advocates have suggested deploying a network of microgrids. Many claim it was microgrids at Buc-ee's convenience and fuel centers and HEB Groceries that saved Houston from the Katrina-like event that was Hurricane Harvey in 2017. A microgrid is a small generation source often combined with energy storage with the capability to isolate from the power grid.

Unfortunately, most utilities still view microgrids as an expensive alternative to what they see as their relatively resilient electric system. These utilities tend to view independent microgrids as competitive to their systems. Some even work to make it more difficult to connect to their grid. I have yet to see a microgrid program where a utility says, "We will help you build a microgrid, interconnect it and pay a fair market value for the excess energy generated." Instead, utilities double down on their centralized view of the electric grid, pledging this time to make it more resilient for that next terrible storm event.

In the aftermath of Katrina, Entergy promised to build more generation plants on the New Orleans side of the river to preserve power generation there in the event a major generation went offline across the river. Unfortunately, the centralized nature of the natural gas plant just built in East New Orleans, requiring transmission, proved again to be the weak link as Hurricane Ida knocked down all eight transmission lines that would have connected to the plant. Citizens of New Orleans in the dark for weeks could only look at the plant their electric bills paid for in disbelief, wondering why it could not operate to keep their food from spoiling or reqiring them to travel for miles only to wait in lines for hours to get gasoline for their automobiles. Worse, the prolonged outage that could have been prevented by proper resilience projects and microgrids is responsible for lives lost in hospitals and elderly care facilities without power.

Sometimes, the disaster is the failed investment itself. In 2008, South Carolina Electric & Gas (SCE&G) and the state-owned utility, Santee Cooper, hired Westinghouse Electric Company, the very legacy of the father of our AC grid, to build two new nuclear reactors at the VC Sumner nuclear plant less than an hour's drive northeast of Columbia, SC. The project would add 2000 MW in clean generation capacity to address growing demand and replace older coal plants that would later be decommissioned. The project was initially budgeted at $9.8 billion but inflated sharply to $25 billion eight years after construction started. Under South Carolina's Base Load Review Act, SCE&G was able to apply for nine different electric rate hikes with the Public Service Commission spiking the average customer's power bill by $27 per month.

Ultimately, in March of 2017, Westinghouse filed chapter 11 bankruptcy. Westinghouse walked off the job after years of mismanagement of the project on their part, and a lack of accountability from the utilities. For months, the two utilities deliberated about how they would complete the new reactors. Finally, on August 4, 2017, both utilities decided to abandon the project. The customers of SCE&G, who had paid more and more and more on their utility bills each year to finance the project, will never see the benefit from their enormous expense. When monstrous mistakes are made with our centralized grid, the pain is widespread.

The fallout from the failed nuclear deal still poisons SC politics surrounding energy policy. However, there has been some healing. In 2019, Dominion Energy, headquartered in Richmond, VA, purchased SCE&G, offering ratepayers lower rates to undo a portion of the rate hikes that now offered no return. Santee Cooper likewise was faced with indecision at the State House about whether they should also be sold. Instead, the leadership was reorganized to lead Santee Cooper beyond this failure. In late 2020, two former SCE&G executives pled guilty to federal charges. Most recently, in August 2021, Westinghouse was ordered to pay $21.25 million to low-income ratepayers that helped finance the multi-billion-dollar failure. Westinghouse executives also face federal charges. But, is anything being done in SC to decentralize the grid? Not much.

My son, Tripp, who has helped me write this book, equates the problem with our grid with one of the very first tactical lessons every private and lieutenant in the army learns. When maneuvering, it is important to maintain proper spacing from other elements on the battlefield. In a foot patrol or vehicle convoy, for example, an IED, RPG, or mortar attack will have the smallest impact on the unit if the spacing is maintained. If the unit is bunched up together, an enemy attack could be catastrophic. The same is true of our grid. Why, then, do we consolidate our critical infrastructure, leaving it vulnerable to the attacks of nature and human error?

The answer goes back to the inefficiencies of monopoly power. A common characteristic of regulated utilities is a low load factor. Load factor is a word that while it may sound technical, it simply compares the capacity to produce power with the actual need for power. A crude illustration of this would be a large 100 megawatt plant. Provided the staff and fuel, the plant can hypothetically deliver 100 MW continuously, 24 hours per day, seven days per week. But, the load on the system only reaches 100 MW for a few hours of the day in a few months of the year (for example in the hot months in the afternoon hours when it's still during the workday and AC units are running). Maybe only a 50 MW plant is needed for the base load – the load that is *always* demanded – and natural gas co-generation plants, solar, and wind that can be dispatched quickly and efficiently can serve the remaining 50 MW. Monopolies are not concerned with efficiency, however.

I faced this issue on one of my own electric bills.

In October 2017, I received a notice from the electric utility, Emera Maine, that serves power to Dandylion House, an old clapboard island house my sister and brother share with our families. The house is situated in the center of the village of Islesford, Maine on Little Cranberry Island, just off of Mount Desert Island and Acadia National Park. Since my three kids and four grandchildren represent the largest segment of the family using the house, I have taken over the responsibility for the electric bills.

The utility was giving me notice, as one of Emera Maine's customers, that they were seeking a rate increase of 5% to cover distribution costs. This 5% increase raised Maine's average electric rate across all rate categories, already among some of the highest in the continental US, to nearly $0.155 (or 15.5 cents) per kWh.

I decided to investigate.

According to the U.S. Energy Information Administration, the overall peak capacity in 2015 for the state of Maine was 4615 MW, and the overall electricity sales for 2015 were 11,741,265 MWh. If the capacity were utilized to its fullest potential (that is, operated 24 hours per day, 7 days per week, and 52 weeks per year) electricity generated would be 40,427,400 MWh. Yet Emera Maine only needed to sell 11,741,265 MWh in 2015. That means that Emera Maine is utilizing a paltry 29% of the full capacity that its ratepayers have paid for. The energy industry would call this a 29% load factor. Furthermore, that percentage is dropping, since according again to EIA, 2016 electricity sales in Maine fell to 11,408,000 MWh; a 28% load factor.

One should think of the load factor as a kind of efficiency or utilization rate for the electrical assets paid for by ratepayers when they pay for electricity. If the load factor is low, it means that the utility has more capital equipment than it needs to serve the customer load. Remember the multi-billion-dollar nuclear project in South Carolina … it still costs ratepayers billions of dollars regardless of whether it generates power 100% of the time, only 28% of the time, or not at all as in South Carolina's case. As a result, electric rates tend to be higher for utilities with a low load factor. Unfortunately, regulators are not paying much attention to load factor, perhaps because they don't understand it, or are being guided by the electric utilities that perform according to the monopolistic forces that influence them. Maybe, it's a combination of both.

I have been researching this issue ever since results in Vermont were published over twenty years ago. The utility worked over those 20 years to systematically raise their load factor from 55% to 70% and lower their electric rates by 4%. Thinking that regulators would start tracking load factor, I began to chart load factors in various southeastern states which continue to be heavily regulated. That's when I really began to understand that regulators were heavily influenced by the utilities they regulated. I knew the public did not understand these issues enough to make regulators and the utilities they regulate accountable.

To clarify, there is nothing nefarious about monopoly utilities acting like monopolies. It is simply their nature as a profit-seeking firm (again, not nefarious, just an economic term for "a business"). There is little incentive for utilities to pursue system efficiency, such as increasing their load factor. There *is* an incentive for them to provide reliable power. On average, citing outage and reliability data from earlier in this chapter, utilities are incredibly adept at keeping the lights on. What I am saying is that maintaining monopoly power in the utility sector is no longer advantageous. Interestingly, we have seen some monopoly utilities, like in Vermont break that mold to their great credit.

Cost of the Status Quo: Oil

The petroleum side of the industry shows a different landscape than the utility industry. Instead of outages and poor load factor, we experience price volatility and sometimes supply disruptions. Our problems with petroleum are largely rooted at the beginning of the supply chain: oil production. Oil production is the source of volatility at the pump because of the fact that 36% of the oil produced in the world in 2020 came from the Organization of the Petroleum Exporting Countries (OPEC) while they also maintained 79.4% of the world's oil reserves.[3] OPEC is a cartel of 13 oil-producing countries like Saudi Arabia, Iran, Iraq, Angola, Nigeria, and even Venezuela, to name a few. Even as the United States has held the position as the world's top crude oil producer from 2018 to 2020, and regardless of whether the US is self-sufficient, such a large single mover in the market on the global scale impacts your wallet.

In 2015, we all saw oil prices drop considerably. Previously, parts of the country had seen gasoline prices higher than $4.00 per gallon. When prices bottomed out in late 2015 and early 2016, some states saw prices as low as $1.70 per gallon. The reason for this was the global supply of oil had grown considerably while demand remained stagnant. Prices dropped again and even went lower because of how the emergence of COVID shocked demand in the first half of 2020. The low prices were quite welcome by most people at the pump. But this caused real problems for smaller oil companies and companies that extract oil utilizing newer technology or methods such as fracking. Many of these companies were American companies and they produced oil in our own backyard. When the price that they could receive for a barrel was less than the cost to produce it, their profits disappeared and many began to go out of business while OPEC gained global market share. The fewer barrels of oil the US and other countries produced, the more that would be purchased from OPEC. OPEC member countries, led by their biggest producer, Saudi Arabia, were losing money with the low prices, too. But they continued to pour oil into the glut to keep growing their market share. As of the writing of this book, the glut has ended and prices are back up close to their all-time highs.

For the consumer, low prices are great! So, why are we even talking about a glut? We aren't talking about prices as much as we are talking about the geopolitical forces that manipulate prices. It could just as easily go the other way ... and it has. When it comes to oil, the 1970s were characterized by volatility and crisis. First, in 1973, Saudi Arabia and other Persian Gulf countries enacted an embargo on countries like the US that supported Israel in the Yom Kippur War. This caused oil prices to more than triple. In 1979, another supply shock would cause prices to shoot up. Before 1979, Iran was

---

[3] https://www.opec.org/opec_web/en/data_graphs/330.htm

one of the top producers of crude oil. However, the Iranian Revolution suddenly cut their supply of oil to the world almost completely off. World oil prices doubled. Whatever you might think about the politics of oil, one thing can be said for certain, letting someone else hold the reins of our oil supply is not in our best interest. When I say "our best interest" I don't just mean the United States as a country. I mean you and me personally.

A complication lurking in the background of all this is the fact that fossil fuels, such as oil, are finite resources. It takes millions of years for organic material to decay and become crude oil. A renewable resource, by comparison, is essentially infinite, or at least replaceable in a very short period of time. So, if something doesn't change, we are going to run out of oil someday. Or, we won't really *completely* run out. But we will start to run out of the easy ways to reach oil, the oil that is cheapest to get. Soon, our supplies will run low, causing prices to go up, which will finally make it profitable enough to pursue the harder-to-reach oil. And so on. Eventually, supply will drag and prices will soar until one of three things happens: we have developed a way to speed the decomposition of organic material by millions of years, our economy collapses, or we have a viable replacement for oil. The challenge here is that we don't really have a way to know when exactly we will run out. In his book, *Out of Gas*, Amory B. Lovins says that we will only see these things happen in our rear-view mirror. Will we be ready or will it destroy us?

Politically, we as a country have responded to all of this in one of three ways: diplomatically or otherwise ensure the supply of foreign oil, foster the production of domestic oil, and encourage the development of technology that frees us from oil entirely. The position of this book is not to prescribe which combination of policy is best, but to highlight the issue of how politics at the international level can affect our lives locally and to understand that we have the power to do something about it for ourselves.

Broader Societal Costs

Back in January of 2020, then Director of National Intelligence, Daniel Coats, warned the Senate Select Committee on Intelligence of the capabilities of China and Russia to cause "localized, temporary, disruptive effects on critical infrastructure." Both our grid and oil & gas infrastructure are considered critical. According to Coats, these effects could be achieved by cyberattacks.

In fact, we saw a cyberattack in May 2021 on the Colonial Pipeline, the largest fuel pipeline in the country. A ransomware attack brought pipeline operations to a halt affecting 5500 miles of fuel transmission. Thankfully, the affects at the pump were short-lived as the company paid the ransom in Bitcoin to get operations going again quickly. The FBI was even able to recover $2.3 million, roughly half of what Colonial Pipeline originally paid in ransom. Where will they get the other half of the ransom? How will they pay for the cybersecurity that will prevent the next attack? It will come out of your pocket every time you fill your tank with gas.

The Colonial Pipeline attack may or may not have been a state-sanctioned attack. And its intent was apparently not to cause harm to American infrastructure but to collect a ransom. Nevertheless, cybersecurity is yet another cost we must bear for the status quo.

A more controversial societal cost is climate change, a politically polarized topic with extremes on both sides. One side rejects the quite obvious reality of climate change as a chronic problem that humanity has caused through emissions of combusted fossil fuels, even calling it a hoax. The other side is convinced the world will end tomorrow if certain governmental policies are not passed. This book isn't concerned about the politics of climate change, so we will commit only to what we know. We know that carbon dioxide, among other gaseous compounds, helps our atmosphere to retain heat from the sun. We also know that burning carbon-based materials like oil, coal, and natural gas releases carbon dioxide. More miles driven in gasoline and diesel-powered vehicles and more kilowatt-hours generated by coal-fired plants mean more carbon dioxide in the atmosphere. We have already put too much in the atmosphere to maintain our current climate as shown by the Keeling Curve[4] (a measurement of CO_2 concentrations started in 1958 in Hawaii). The Keeling Curve shows that concentrations have risen from 280 parts per million volume (PPMV) to nearly 420 now, a level not seen in over four million years.[5]

In just the last couple of decades, I have watched an old piling of stones from what used to be a dock in the Islesford, ME harbor dip further beneath the surface at high tide. It wasn't 20 years ago when the top stones never went underwater. Today, every high tide erases the pilings from sight. Rising temperatures have slowly melted ice caps, increasing the volume of our oceans. How much longer before flood maps are redrawn and more homes in cities like Charleston, SC or Houston, TX must buy flood insurance? Hurricanes, too, are getting stronger and more frequent as the average temperature of the water that feeds them slowly rises. On the west coast, already dry climates are at increased risk of wildfires.

---

[4] https://keelingcurve.ucsd.edu/
[5] https://theconversation.com/climate-explained-what-the-world-was-like-the-last-time-carbon-dioxide-levels-were-at-400ppm-141784

No, climate change may not be threatening to end the world. However, it is certainly getting more expensive to live here.

A Challenge

Dream with me for a minute. Imagine how life would be if all of the problems we just talked about disappeared.

What if we never had to fill another tank of gasoline?

What if we could use whatever energy we wanted because _we_ produced it?

What if we turned energy supply from a liability into an asset?

That's the challenge I am asking you to take: be your own energy supplier. Seriously. *It really is that simple!* Yes, it is going to take commitment and it will take some time. The good news is that anyone can do it, especially you, because you're reading this book and understand why it must be done. Your family's energy security depends on it.

We are investing in our energy future, and this book will show you how. Every dollar you spend on energy improvements will incrementally reduce the money you pour into the status quo. Conversely, the coolest thing about taking charge of your energy is that every dollar you keep from pouring into the status quo is a dollar used to improve your own energy supply. That next improvement will create even more savings and drive you even further down the road to Energy Freedom.

The best part about working toward Energy Freedom is that there are no big upfront investments, no debt, and no negative impact on your budget. It will start with a crawl, then a walk; before too long we will be running on our own energy.

Let's get started!

3

The Energy Freedom Fund

The average family of four consumes as much as $1 per hour, on average, in energy expenses. This amount is given to the electric or natural gas utility, fuel oil, or kerosene dealer and gasoline station. Once the money is spent and the energy is consumed, it is gone forever. However, the utility and oil industries will reinvest that $1 per hour that you pay into infrastructure so they can continue providing energy at the status quo to continually collect $1 every hour from you. That amounts to $8,760 per year. If instead, you were to invest that money from age 30, it would turn into $1.58 million by age 60 (assuming an annual return of 10.5% – about the average S&P 500 return since 1926). But you need the energy for transportation and to keep the lights on. So, one hour from now, you will hand over another dollar. And then another.

In this book, we will tell you how you can break that cycle and invest most of that dollar in your own energy infrastructure. Pursuing Energy Freedom is sort of like becoming an energy millionaire. And anyone can do it by following the steps we have laid out in this book.

The 30-year opportunity cost of doing nothing is $1.58 million.

This all started for me in 1997, when, as part of a career adjustment, I felt strongly that I needed to put my own energy ideas to work in our own home if I were going to be successful at telling others how much they could save. I had 15 years' experience at Georgia Power, Sylvania Lighting, and EPA Energy Star as well as Solarex Solar Panels. By then, I believed we could achieve net-zero energy in the home, but could we do it as a strategy to compete for investment return as I would expect in other funds? So, I put in some energy-efficient light bulbs, wrapped our water heater, and put it on a timer to start. By the end of 1998, we had reduced our electric bill from $145 per month to around $106 per month.

Then in 1999, a science fair project with my then 12-year-old son and contributor to this book provided the opportunity to start measuring prospective energy savings for major appliances. Using the electric meter and turning on and off different appliances we were able to extrapolate the energy they used. For the new appliances, we would use the Energy Guide "Yellow" tag that predicted an appliance's energy consumption to compare to our own appliance to see how much we could save.

I promise you, we will not be reading meters.

Personally, I track my energy savings with a high level of detail, because I want to know exactly where we stand. It helps me to tell my story with some real numbers, which I will do whenever relevant. But let's keep things simple.

Our plan will be to use energy savings to reinvest into the next energy upgrade (Figure 3) increasing the savings to a point where within six to ten years (if you follow the plan aggressively) you can reduce that dollar spend to $0.20 per hour saving up to 80%! The reinvestment of savings not only begets more savings but builds up your own energy delivery system. Just as one would rather own their home to build up equity rather than rent, I would recommend taking this opportunity toward Energy Freedom.

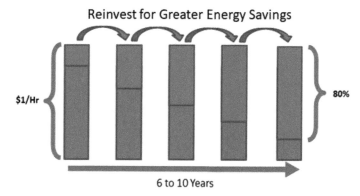

Figure 3.   The 6–10 year energy savings reinvestment plan.

The very first thing we must do is establish your energy baseline. Your energy baseline is your current situation, your starting point. How much energy do you use in a year? How many gallons of gas do you burn in a year? We need to know this primarily so that we can measure your savings over time. When your energy costs are decreased through energy improvements, use the savings to fund the next improvement. But we have to know your energy baseline before we can calculate savings.

Your electricity baseline is pretty easy to figure out. You can get the previous 12 months of your power bills from your utility and calculate your average monthly cost. It's important to look at an entire year as the seasons have an impact on your energy usage. Make sure not to count any costs other than the cost for energy consumed (in kilowatt-hours or kWh). Your bill will include basic, fixed charges that will not change. When we save energy, we save money only on the cost of energy we no longer have to purchase. You may also want to know your average consumption in kWh as well as your average rate (the cost per kWh). If you have a flat retail rate, the cost per kWh doesn't change. But sometimes electric rates are different for summer months or winter months. Rates may also be tiered for different levels of demand (i.e., over 800 kWh usage on a bill the rate goes up or down). With the average monthly consumption and cost per kWh, we can get additional insight into how energy upgrades save money.

Your monthly gasoline consumption, on the other hand, will have to be estimated, unless you are incredibly nerdy and document your mileage and cost of fill-ups as I do. If you're perfectly normal, then we'll estimate. You can do this in three steps:

1. Figure Average Monthly Mileage: If you don't know how many miles per year you drive, take the total number of miles you have driven with each of your vehicles and divide that by the number of years you have owned the vehicle. Adjust for years when you drove an unusual amount of miles, for example, a one-time vacation that was farther away. Once you have an average annual mileage on each vehicle, divide that number by 12 for average monthly mileage.

2. Figure Your Fuel Economy (miles per gallon or MPG): Most cars manufactured in the last couple of decades or so may have the MPG rating already published with the car's materials or online at www.epa/fuelconomy.gov. Worst case scenario, you can calculate this manually. First, set your trip odometer at zero and monitor your fuel consumption for a month. Also track the number of gallons of gas consumed (not the number you purchased – you may still have a full, unused tank at the end of the month that would throw off your calculation). Then take the number of miles from your trip odometer and divide that number by the gallons of gas you used in the same period. Your fuel economy is different on the highway versus in town, in the heat and in lower temperatures, etc. Try to get a mixture of circumstances in your driving to get an accurate MPG estimate if calculating manually.

3. Figure Your Monthly Fuel Cost: With mileage and MPG we simply take the average monthly miles driven and divide by the MPG to give us the number of gallons of gas we purchase in a month. Finally, multiply that

by the typical price per gallon that you pay. This final step can be constantly adjusted as gas prices fluctuate over time.

If you are curious, compare your averages with the US average. The average US electricity bill was $115.49 per month in 2019. Keep in mind, this was only for electricity (kWh) and excludes any natural gas or fixed charges that you may see on your bill. Average gasoline costs are much more subjective. If we assume the national average of 14,300 miles, a two-car household, an average MPG of 25, and the average price per gallon as of August, 2021 of $3.25, the average monthly household cost for gas would be $309.83.

Although I want you to keep track, going forward, of your gasoline consumption and your vehicle's fuel economy, for now, we are more immediately concerned about the utility portion of the baseline since we will be chasing utility bill savings first. They tend to be the lowest cost, easiest return on investment, whereas, there's not much reducing fuel consumption in any meaningful way without straight-up purchasing a hybrid-electric or electric vehicle (EV). We're not there yet unless of course, you were about to go out and replace a vehicle this month. If so, please skip ahead to Chapter 6 and read about replacing your vehicle. It may still be a few years from cash flowing an electric vehicle from your energy savings, but you may consider it if you are replacing a vehicle sooner.

STEP 1: Start with $50

The way to be sure to acheive Energy Freedom is to have a separate account that savings will flow into and out of which energy upgrades will be paid for. We will call this the Energy Freedom Fund. It can be a simple savings account that you transfer your monthly energy savings to each month. Knowing your baseline, each time you make an upgrade and knock a few dollars off of you monthly energy costs, add those dollars to your monthly transfer to the fund.

In order to start, we need to inject something into our Energy Freedom Fund before it starts to move on its own steam. Imagine a locomotive at the top of a hill with a shallow grade. It's going to require some nudging before it starts to move. Your nudge to get this train rolling is just 50 dollars per month. And, that's all you are going to need for your first energy improvement anyway. That small investment will start generating some small returns. Capture those returns the following month and reinvest them, adding them to the next 50 dollars, and watch the fund grow as you continue with $50 per month but adding in additional energy savings.

Now the railroad track's slope is gradual, so it'll take a few more nudges before gravity causes the locomotive to accelerate. At some point, our nudges will mean less as it eventually starts running faster and faster on more of its own steam. That's when we jump on and enjoy the ride to Energy Freedom. But, if you want to get there faster, keep nudging.

Depending on your situation, make your nudges bigger to accelerate your locomotive more quickly. You may also want to consider how your energy expenses compare to the average. A higher than average energy expense is a bigger locomotive and may warrant some bigger nudges. But, $50 a month is something virtually anyone can do and is enough to get the wheels turning right away for the average energy user. Make this automatic with scheduled transfers. Spread it out, if you want, to $25 twice a month, or $12 every week.

The good news is $50 is more than enough to begin immediately on the small, inexpensive improvements. As we progress, we are going to use those initial savings to chase bigger and bigger improvements. Larger energy upgrades that will ultimately provide larger savings will require a larger investment. Continue setting aside $50 per month plus your savings from previous improvements until you have the cash for the next one. I recommend using a separate savings account or money market account that is easy to get money out of when the time comes. I do not recommend putting the money anywhere else. CDs will tie the money up for months or even years without a return to make it even worth it. Investing in equities is riskiest when money is invested for less than five years or so. Your next energy improvement will not take nearly that long. Separate your Energy Freedom account from your day-to-day checking account so you can't easily dip into it for pizza on Friday night and set yourself back.

Investing in Energy

Let's admit it. We're making an investment in our energy future. Just as we invest in 401(k)'s, IRAs, etc., for financial freedom, we are investing in energy improvements to achieve Energy Freedom. So, we treat the dollars we put toward energy improvements the same way we treat the dollars we put toward our financial investments. When we purchase lower wattage energy-efficient light bulbs for example, we should consider their cost as the investment and their energy savings as their return on investment (ROI). A dollar saved from an avoided energy purchase from the electric grid (or the fuel pump for that matter) that we no longer need thanks to the energy-saving light-bulb shows up as a reduction against our energy baseline. Thus, our returns are measured in how many new dollar reductions from our baseline are occurring versus the dollars we spent to receive them.

Here is where it gets interesting. Think of all the investments in the financial world. Your return there is measured in the investment income that comes back to you by way of rent income, capital gains, interest income, etc.

However, as income, your returns are taxable. Additionally, trading equities, buying and selling investment real estate, and just about any other financial move come with commissions and fees above the taxes you will pay. But energy savings returns are pure, free of taxes, commissions, and fees. Other than the cost of the improvement, the savings are 100% yours.

Consider an investment in equities ... say, the S&P 500. Over the long term, the S&P averages annual returns of 10.5% per year. Assuming your money is invested in a 401(k), all growth is taxable at distribution as ordinary income, say at 25% for federal taxes and 7% for the state. On top of that, you have been paying 3% in commission or fund management fees each year. Your net return is really about 7.1%. So, comparing this investment to investments in energy improvements, we should be willing to accept energy improvements with ROIs at 7% or more.

We're going to go through each energy improvement in detail that will bring returns to your Energy Freedom Fund. For now, let's peek at what your investment will look like (Figure 4).

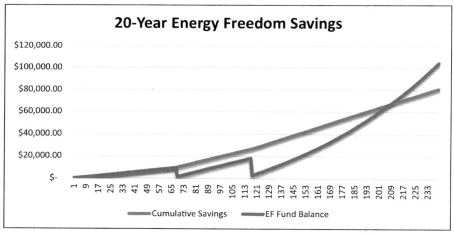

Figure 4.   A comparison of cumulative savings to the overall amount achieved by investing the savings after achieving energy freedom (by year 20).

With our $50 per month contribution, we will be able to chase quick improvements for quick savings. Those savings will be added to our monthly contributions which will enable us to make the next upgrades sooner, begetting more savings. Thus, our energy savings will compound just like traditional retirement savings.

In just 20 years of contributing $50 per month and reinvesting our energy savings, we achieve an average annual return of 12.12% yielding over $50,000 in our Energy Freedom fund. If we had put that same $50 in an S&P

500 index fund for 20 years at its average 10.5% annual return, we would only have $36,378. But, if we invest our savings into an S&P 500 index fund after investing in solar energy, we can get an average annual return of 16.26% yielding $90,000! And, there are no taxes or management fees on your Energy Freedom investments.

Return on Investment for Energy Improvements

The ROI is the basis for charting how we are doing and in what order we implement the technologies to get the train rolling faster. All you need is a simple ROI calculation. And all that requires is the cost of the improvement and the annual energy savings from the improvement. ROI is simply annual savings divided by cost. The cost part is easy ... look at your receipt, purchase order, contract, or whatever shows the cost. The energy savings may require some estimating.

Installing a single LED bulb is the tiniest, most incremental, and most affordable improvement you can possibly make. It is also the highest ROI improvement you can make. That is why we will start there. We will get more into the specifics of a full lighting upgrade in the next chapter. But you calculate the ROI the same with an LED upgrade as you do with just about any other energy upgrade. And there will be many different energy upgrades we will be working toward. In fact, each upgrade will have its own unique cost, savings, break-even point, and overall ROI. But the lighting upgrade provides an easy ROI calculation example.

A typical 9 W LED light bulb will likely cost two dollars per bulb. However, when purchased in larger boxes, typically 24 bulbs, they can be as cheap as $1.50 per bulb. The energy savings, however, will depend on a few factors. First, we need to know what bulb the 9 W LED bulb is replacing. Whether it is replacing a common, but costly, 75 W incandescent bulb or a more efficient 23 W compact fluorescent bulb will impact savings. Second, we need to know how often the bulb is used. If we are replacing a bulb in a closet that we only turn on for a few minutes each week, we won't save as much compared to a bulb in the kitchen that shines for hours every day. Lastly, we need to know our electric rate. If we avoid buying a kilowatt-hour from the utility by installing a more efficient light bulb, we want to know how much that kilowatt-hour would have cost.

Let's make some assumptions: we are replacing a 75 W incandescent bulb that is used four hours per day, on average, and our electric rate is ten cents per kWh. We are reducing the wattage of that light by 66 watts. A kilowatt-hour is a unit of energy whereby we use a kilowatt (1000 watts) for one hour. So, in one hour of using our new bulb, we save 66 watt-hours, or .066 kWh, since that is how we are billed. At four hours per day, we save 96.4 kWh per year, or $9.64. Dividing our savings by our cost of two dollars for the bulb, our return is 482%!

Some technologies have vastly different ROIs, many of them much lower than 482%, of course, but are considered integral to Energy Freedom. For example, a solar PV system, discussed in Chapter 7, will provide a comparatively small ROI, even among the lowest you will pursue. But the investment is still worth pursuing and we will need our own generation in order to make Energy Freedom possible.

And there are many improvements in between with varying ROIs. These energy-saving opportunities should be ranked such that we can follow them in order, always chasing the highest return first. If we pick a low ROI improvement out of order over a higher ROI improvement, the lower return will act as a brake on our momentum slowing down the rate at which we accumulate cash savings for our next investment. Therefore, we will order them so that the highest return improvements quickly feed into the next highest return improvement. Following the reinvestment plan laid out in this book will ensure the highest possible combined Energy Freedom ROI. Therefore, I recommend you might follow a ranking such as the following (Figure 5.):

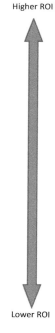

Higher ROI

1. LED Lamps (replacing incandescent bulb)
2. LED Lamps (replacing compact fluorescent bulb)
3. Insulation for Water Heater
4. Timer for Water Heater
5. Programmable Thermostat for Heating and Cooling
6. Weather Stripping Doors and Windows, Outlet and Switch Plate Covers
7. Toaster/Microwave Oven
8. Energy Star Front Load Washer and Dryer
9. Energy Star Dishwasher
10. Heat Pump Water Heater
11. Energy Star Refrigerator
12. Plug-in Hybrid Automobile
13. Foam or Cellulose Insulation in the Attic
14. Electric Vehicle
15. New Heat Pump A/C System
16. Solar PV Roof

Lower ROI

Figure 5.   A ranking of energy measures by their return on investment.

For some of the larger improvements, we may only want to count the incremental cost of the improvement over the incumbent technology we are replacing. For example, if we are talking about the electric vehicle, don't count the entire MSRP of the vehicle in your ROI calculation. You were going to own a vehicle anyway. Rather, consider the cost of the electric vehicle over the cost of the next vehicle you would have purchased if you had never read this book. Yes, your ROI will appear more attractive by calculating it this way. But, this does show the real cost of the improvement more accurately

We are building a portfolio of energy investments. As such, we should think of each energy improvement as a part of the overall portfolio. The performance of the portfolio is greater, on average, than the performance of individual energy investments because the portfolio reduces risk. At some point, we are going to be investing in energy generation. But, before we get there, we have to optimize the way we consume energy; we have to become more efficient. If we don't, we will be forced later on to generate more energy than necessary. That means a larger energy generator, such as a solar energy system, and a much larger investment in order to power our inefficient consumption. Installing solar is a great investment and a crucial part of your energy portfolio. But, buying a larger solar energy system so that you can power your incandescent light bulbs doesn't make any sense and will hurt your portfolio.

You can think of your energy portfolio in much the same way that you think of your stock or mutual fund portfolio. Diversification reduces risk. The difference is where that risk comes from. In the financial world, the future of equities is largely uncertain due to systematic and unsystematic risk. Our confidence rests mainly on past performance and the belief that future performance will be similar. So, we hold many different types of stocks (or mutual funds which diversify for us) to mitigate the risk in one stock or in one industry group of stocks to our portfolio.

Your energy portfolio reduces risk in a different way, however. There is little risk in the individual investments themselves. If you install a 5 kilowatt (kW) solar energy system, you are going to generate 5 kW of power. You don't need a crystal ball to predict the future performance of your energy investments. Though you do need to plan out your improvements so that every individual investment is optimized within the portfolio. Thus, the risk in your energy portfolio grows from disorganized planning of energy investments, such as installing solar before reducing your consumption through efficiency.

When all is said and done, you will have a healthy portfolio of well-planned energy improvements that combines the low cost, high ROI of lighting and appliance upgrades with the larger cost and higher density savings, but lower

ROI of improvements such as solar and electric vehicles. The result is a system of energy upgrades that work together in synergy where the ROI of the portfolio is greater than the weighted average ROI of the individual improvements.

Final Words before Our First Investment

All of our effort in the beginning of this journey is going to be focused on energy efficiency. For some folks, the phrase "energy efficiency" may conjure up some negative feelings that efficiency means austerity. You may be thinking that to achieve Energy Freedom, you have to take colder showers, use dimmer light bulbs, and hang your clothes out on a line. But none of that is the case. Energy Freedom is about finally having the liberty to consume whatever you want because you own it. It starts with becoming more efficient, but it ends with you creating and owning all of your energy.

Do with *your* energy whatever you desire.

Caution: You will be tempted to make some small initial energy improvements, experience some savings, and then quit working toward Energy Freedom. Maybe you are using those savings to beef up your retirement accounts; maybe you are using them to increase your lifestyle a bit. Whatever the case may be, there is so much more to be had. Don't sell yourself short!

If Energy Freedom is like a locomotive, then perhaps monthly energy savings are like a speedometer. A train that doesn't run, isn't much of a train. But, when it's running, knowing how fast our locomotive gets from point A to point B tells us how valuable it is. A faster train moves freight faster and gets paid sooner. Such is the case with Energy Freedom. This low-hanging fruit injects quick savings into our Energy Freedom which makes the train move and helps us chase new energy improvements.

$50

4

Low Hanging Fruit

When you first begin investing financially, the returns are very slow to begin with. You put in a thousand bucks to start out, and although you expect ten percent annual returns on average, ten percent of a thousand is not very much. You continue to invest bit by bit and the balance slowly increases until some point far in the future when the rate of return starts to make a real difference. It takes a long time before your money starts earning more money on its own in any meaningful way. That's not to discount the power of compounded returns. It just takes a while before that power really comes to fruition.

Not for us. We are going to shoot out of the gate like a lightning bolt. As mentioned previously, one of the first investments we are going to make is incredibly affordable and offers returns up to and even beyond 800%, very quickly adding to our fifty-dollar monthly energy investment budget. Figure 6 shows how quickly we anticipate building energy savings in the first 12 months.

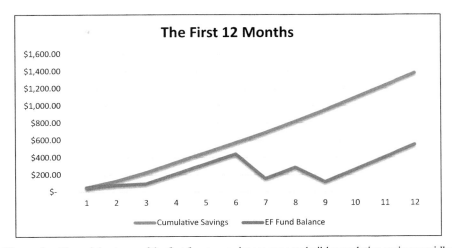

Figure 6. The quick returns of the first few energy improvements build cumulative savings rapidly.

There are even some free things you can do to harness some savings. In that sense, the locomotive analogy doesn't quite do justice to these first energy improvements. The locomotive is more of an analogy for your overall journey toward Energy Freedom. But our energy investing in the very beginning is going to work more like a rocket: accelerating rapidly away from the earth and only reaching a plateau in velocity once it escapes Earth's gravitational pull. These first energy improvements are the rocket fuel we need to take off.

That's the goal when starting out with our energy investments. We want to get as many savings as possible upfront so that we see savings sooner and we build some serious momentum before tackling the larger investments later on. In plainer terms, we should be able to triple our monthly contributions to $150 in a few short months.

STEP 2: Find the Free Savings

Free is pretty awesome, but it's even better when free generates a return. If you can save a dollar without investing a single penny, then you have achieved the elusive infinite return. That doesn't even exist in most places. But you can easily get an infinite ROI by simply making some small changes in the way you use energy. Changing our habits and our behavior around our energy use costs us nothing, and it can save a few bucks a month.

Yes, we are on the road to Energy Freedom, not energy austerity. So, these free improvements are entirely up to you to pursue. Changing any habit takes a little work and discipline at first, but you don't have to change your lifestyle to the point of inconvenience just to save a buck on your monthly energy bill. Early in my family's journey to Energy Freedom (back when it was really still just MY journey) I told my wife she should air-dry her hair instead of using her nice hairdryer. She responded with something along the lines of "like hell."

Do what makes sense for you and your family. But, remember: the more you are able to achieve in this second step, the more rocket fuel you start with.

Here are some ideas:

Heating and Cooling:

For just about any home, heating and cooling loads are the largest. Conditioning the inside temperature to be consistently 20 or more degrees warmer or cooler than the outside is no small thing. This is especially true in the Southeast which boasts some of the lower electric rates in the country, but some of the highest average utility bills. So, there is more potential for energy savings by adjusting the way you heat and cool your home than by any other energy savings effort. But we often take for granted just how much energy is required to make that happen. For most climates in the US, you may notice that your energy bills in the spring and fall months are the lowest in the year. That's because the temperature outside is mild and not much different than how you would like the inside of your home to feel.

For the rest of the year, though, you can make a big difference by letting your home get two degrees warmer than your ideal temperature in the summer and two degrees cooler in the winter. For example, if you feel perfectly comfortable at 70 degrees, set your thermostat to only cool the house when it gets above 72 degrees and to only heat the house when it gets below 68 degrees. Your HVAC system has to work harder and harder for each degree it has to heat or cool away from the outside temperature. In other words, your system will work harder to go from 71 to 70 degrees than it will to go from 81 to 80. That's because the outside temperature will be warmer than 70 for more time than it will be warmer than 80 during the warmer months of the year. When the temperature rises in the morning on a hot day, it passes 70 first, then 80, peaks, and then goes back down in the evening passing 80 and finally 70. The same is true in inverse on a cold day. Furthermore, if the high temperature is 80 degrees outside, 72 degrees inside is closer to the outside temperature than 70. Your HVAC unit will work harder to cool or heat for each degree the further away the inside temperature is from the outside. By letting your home get two degrees warmer in the summer or cooler in the winter, you have relieved your HVAC system of the burden of conditioning those last two degrees.

During the winter months, if you have electric heat, utilizing a fireplace to heat your home can be a great energy saver. A natural gas-burning fireplace insert is great for this. Personally, I prefer a wood-burning fireplace just for the environment it provides, so that is fine if you prefer. Heating with gas is currently less expensive than heating via electricity unless you have an electric heat pump. Because I am concerned about climate change, I signed up for renewable natural gas which I get at far more competitive costs than electricity. So, if that option is available, use it first before turning to the central heating system. To heat the entire home with a single fireplace, turn on your central HVAC system's fan. By running only the fan, you circulate the air, moving the heat from the fireplace to other rooms without using the electric heating element in the HVAC system as much. This is most effective in smaller homes.

While using both gas and electricity for heat, keep track of the cost of gas versus the cost of electricity. The great benefit of having both options is that you can select the cheaper option to heat your home. If the price of natural gas spikes, you can revert back to electric heat. When the price is low, as it has been maximize the use of gas for heat.

You should also consider how radiant heat can affect the heating and cooling of your home. Radiant heat is essentially heat generated in the infrared spectrum and accompanies the energy from light sources such as the sun.

The infrared energy coming in through your windows from the sun is actually heating your home whether you want it to or not. In the summer, this can create unwanted heat gains and cause your HVAC system to work overtime in order to cool the house. Consider shading your windows with insulating blinds or drapes. In the winter, take advantage of the radiant heat gain. Keep the blinds up and the drapes drawn during a sunny day to let in the light and to take advantage of this free heat. Your cat may even thank you.

Cooking:

Depending on how much you cook at home, there are very likely some savings that have been left out in your kitchen. Similar to the HVAC discussion, your refrigerator and freezer can be set so that your food stays cold while using less energy. Simply turning your Freezer up to 4 °F (−15.6 °C) and your refrigerator up to 40 °F (4.4 °C) is sufficient to keep food safely frozen or cold. While the FDA does recommend your freezer be set to 0 degrees, their recommendation is only necessary for long-term storage of food as 0 degrees will make food safe indefinitely. Most families cook frozen food within a week or two of purchase. So, those few degrees that your refrigerator and freezer *don't* have to work through, saves some energy.

Choosing low-energy use cooking methods also can have a big effect on your kitchen energy use. Toasting a loaf of Italian bread in a toaster oven instead of your conventional oven will yield the exact same result (seriously, there will be no difference in the quality of your crunchy, yet soft and buttery Italian bread). And, according to an Energy Star Market & Industry Scoping Report on toaster ovens, you can save one third to half of the energy it would have taken in a conventional oven to cook a small-to-medium size dish if you cooked it in a toaster oven. The main reason for this is the difference in size between the toaster oven and the conventional oven. A conventional oven has to heat a much larger amount of space than the toaster oven does, even though the conventional oven may be better insulated. So, unless you need the space for a large dish like a ham, a cake, or multiple dishes, use a toaster oven if you happen to have one.

Using the microwave to augment your cooking is also an effective way to reduce cooking time in your energy-intensive oven. When baking potatoes, for example, six minutes in the microwave can reduce cooking time in the oven by 20 minutes. Even if you were to bake them in a toaster oven, your microwave gives you a big head start by transferring heat energy to your potatoes faster than any other appliance in your kitchen. Remember: energy equals power multiplied by time. Opting for appliances that use less power is one way to reduce energy consumption. Reducing the amount of time the appliance draws power is the other way. And, compared to an electric oven, a microwave does both.

Hot Water:

For a warm bath, boil water on your gas range to fill your tub. Kidding!

In all seriousness, there are three main uses for hot water: bathing, laundry, and washing dishes. We can probably agree that washing dishes and maybe doing a load of whites are the only times we might want to use really hot water (some dishwashers heat water with their own internal, electric heating element). But, does it really need to be scalding? Would your dishes not be just as clean if the water were as hot as a comfortably hot shower? Try turning the dial down a bit on the water heater to where the hottest your water will get is as hot as you would like your shower. A five-degree difference on a typical electric 60 gallon water heater in the average home could yield as much as $5 per month in savings. If the 60 gallon water heater used gas, you might only save a few cents each month ... but all you did was turn a dial. Why not?

Laundry:

Hot water aside, your clothes dryer can hog a pretty good portion of your overall energy load. We'll discuss energy efficient appliances later on, but the fact remains that blowing hot air on a load of clothes and tumbling them for an hour requires quite a bit of energy. The simple answer to reducing energy consumption while doing your laundry is to reduce the amount of water that makes it into your dryer.

There are a few ways to make this happen. First, be careful about selecting the size of the load during the washing cycle. Telling your washing machine that you have a large load versus a small load is telling it to use more water. I'm not saying you should use less water than you need, but I am saying that your selection matters. That being said, you should always try to run your laundry cycles (and your dishwasher, for that matter) with full loads. If you are only washing two socks, a t-shirt, and a pair of jeans, you are wasting energy.

Second, if you have the option, select an extra spin cycle. Spin cycles use centrifugal force to pull water out of your clothes when the wash is finished. This reduces the amount of water that goes into your dryer by spinning it out beforehand. Yes, an extra spin cycle will use more energy, but it will save you more energy when your clothes are drying. It is important to note that energy-efficient washers use less water and, therefore, do not need the extra spin cycle. It may not even be an option. Play around with the settings on your washer to see what limits dryer time most effectively.

Finally, reducing the articles of clothing you put into your dryer from the wash cycle, obviously reduces the water that makes it into the dryer. Hanging clothes to dry uses absolutely no energy at all and actually makes your clothes last longer. I typically hang some of my nicer, casual shirts and pants rather than put them in the dryer. True, this means that the load I put in the dryer is now not a full load. But, the dryer functions more on how much water it has to remove than it does on running through a full cycle. As long as it is set to run on a setting that senses the moisture in your clothes, it will stop when the job is done.

Carpooling:

If both of you and your spouse work near each other, carpooling in the more efficient automobile and leaving the less efficient vehicle at home can save significantly on gasoline (and wear and tear). We were able to save $600 in a period of eight months of carpooling. Admittedly, carpooling introduces a little more inconvenience than some of the other opportunities for free savings. Feel free to take it or leave it.

These suggestions for free energy improvements are not all-inclusive. Be creative and energy conscious as you look around your house and you will find some inefficiencies to tighten up so you can harvest more free savings. Once we stop looking at energy as a given, this step becomes much easier. If you have an average family with an average home, you can easily find $7 to $10 in savings per month just by adjusting the way you consume energy. By the time you receive your next utility bill, you will have $60 per month going toward future energy improvements.

STEP 3: Upgrade to LED Lighting

There is often a lot of misrepresentation in the marketing of light bulbs. So, before you go out and spend your entire first month's $50 in savings on your lighting upgrade, we need to set the record straight. You'll notice that LED bulbs will show "60 Watt" or maybe "40 Watt" in a conspicuous location on the package in large font with the word "equivalent" in an unnoticeably small font underneath. Lighting marketers assume that the average light bulb purchaser doesn't know what a watt is and thinks it's a unit of light. So, if they show the actual power usage, which could be as low as 8 watts, they fear that their bulb will be confused for a dimmer bulb and be put back on the shelf for the competitor's bulb. The intent of the "60 Watt Equivalent" marking is to convey that the LED bulb emits as much light as the much less efficient 60W incandescent bulb.

Their assumption is probably justified; most people don't know what a watt actually represents. But you do.

Don't be confused by the packaging that attempts to compare the light output with a unit of power. The packaging will show the actual wattage of the bulb in a less obvious place on the package. That's what you are really looking for. However, if you are concerned about light output, look for "lumen" in the specifications on the box and compare it to the bulb you are replacing. Your old bulb may have its lumen output printed on the base of the bulb.

STEP 2 aside, installing an LED bulb is the tiniest, most incremental, and most affordable improvement you can possibly make. It's so affordable, depending on the size of your home and how many light bulbs you have to replace, you may very well be able to upgrade half of your home with the first $50 in savings from the first month. And, at the very least, you should be able to replace about 15 of your most inefficient bulbs all at once.

If you have a lot of bulbs to replace, you will want to prioritize your lighting upgrade by replacing the higher wattage light bulbs first. So, to start, take an inventory of all the light bulbs you have around your house, categorizing them into three groups: incandescents, compact fluorescents, and LEDs. For most light bulbs, you can easily identify what type a bulb is just by looking at it.

Many incandescent bulbs, the most inefficient, have a clear glass bulb. The clear bulb should allow you to see inside the filament, a material strung between two electric nodes that glows when the nodes deliver a current. While incandescent bulbs are very effective at putting out a lot of light, they require a high wattage to make the filament glow. Because of this, they get very hot. That heat represents energy that is wasted since most of it is in the infrared (non-visible region). Because LED technology offers such flexibility in the design of various types of lighting, marketers have designed standard household LED bulbs to look like traditional incandescent bulbs as can be seen in Figure 7. Be sure to verify on the package that you are, in fact, buying LEDs.

Fluorescent bulbs are characterized by a glass tube made opaque by the fluorescent powder that coats the inside of the glass tube. It glows when current

Figure 7. The A-Line LED Lamp looks like a standard incandescent but provides light for 15% of the power.

flows through the electric nodes at the ends of the tube. Compact fluorescents (CFLs) are fluorescent bulbs where the tube is smaller in size and is twisted or folded around into a compact shape so that the bulb can be used in the same sockets in which an incandescent would be used.

LED stands for light-emitting diode. Instead of using a filament or fluorescent powder, LEDs use a diode. The diode is essentially an electronic chip where electrons jump from one point to another. This jump emits light. LEDs often sport the traditional bulb shape characteristic of the traditional incandescent, but they are highly versatile and can fit in many different types of lighting applications. Aesthetically, it's nice to return to the consistency of that traditional bulb shape we are used to, after some of the funnier shapes of CFLs. Another benefit of LEDs is the ability to manipulate the hue of the light output without sacrificing brightness or efficiency. LEDs can be made to shine bright white, with the same quality and hue of light as natural sunlight. This type of light is very popular in retail as it maintains the fidelity of color in the merchandise on display. They can also carry a softer glow more similar to incandescents for the cozier feel in your home you may be used to.

Start by replacing all incandescents and then replace all flourescents next. Naturally, if you already have LEDs in place, there's no need to change them out. Recent efficiency gains in LED technology may not be enough to warrant changing an older LED for a new one.

STEP 4: Improve Water Heater Insulation.

Water heaters easily consume ten percent of the energy use of the typical home costing as much as $10,000 over the course of a 30-year mortgage. We know our homes need hot water for several different reasons. Yet many water heaters are manufactured with minimal insulation and are very often installed in unconditioned areas of homes like a garage or attic. As a result, your water heater has to run more often than it should, and as much as a quarter to a third of energy is therefore wasted. The solution is simple: add insulation. A $25 thermal jacket, or insulation and duct tape, can save five to six dollar per month for an electric water heater. These are easily wrapped around the heater and taped in place and can be installed in less than ten minutes. This generates about a 350% annual return on investment. Energy improvements don't get much easier than that!

As always, once you figure the monthly savings of an energy improvement, add that amount to the monthly savings you have already accumulated. We went over ROI calculations in the last chapter using lighting as an example already. Taking what we figured by now, we have achieved about ten dollars in free energy savings, $34 in savings from your lighting upgrade, and six dollars from wrapping your water heater. Adding that to our $50 monthly contribution, we are now putting $100 into our Energy Freedom fund every month.

And, we're still just scratching the surface. Our future improvements will gradually get more capital intensive, so hold on to that $100 per month, building up the fund for the next one.

$100

5

Finding More Savings

We've accomplished the easiest tasks. Now it's time for us to roll up our sleeves and dig in a bit more to keep this locomotive rolling. It lumbers inch by inch, slowly toward a steady downward slope, but not yet of its own power. There are more nudges yet needed. These next improvements will gradually cost a bit more but they will provide bigger and bigger nudges by adding larger chunks of savings to our monthly contributions. But we are still talking about small improvements that will bring new savings quickly. We started with the free improvements with an infinite ROI. Then we upgraded lighting for up to a 500% or more annual return. As our nudges get bigger, the costs of upgrades will rise, and return will come down as we approach the three upgrades most critical to Energy Freedom: an electric vehicle, a solar PV system, and energy storage.

By accomplishing the upgrades in this chapter, we should be saving $125 per month in energy costs on top of our 50-dollar contributions. Based on Figure 8 showing our experience we will be 25% of the way toward our Energy Freedom goal. That means we will be contributing $2,100 per year to our energy freedom fund. We will start talking in years because it will take longer for the large upgrades. Before that, we must squeeze out the rest of the energy efficiency savings we can get. Based on our experience as a family, having sought to become zero energy, there are limits to how far you can go in saving energy in the home. After a certain point, there aren't any more opportunities for savings that offer an attractive enough return or that benefit the overall Energy Freedom portfolio in a meaningful way. So, the highest average energy savings in the home we should probably expect is $125 per month, unless you are undergoing a major renovation or building new.

To see how your total energy efficiency savings fit into your overall Energy Freedom, take a look at this chart of our savings off of our baseline. Energy efficiency, which we will complete in this chapter, is the foundation of Energy Freedom. But there is much more to pursue. At only 25% of our total energy cost savings, we save $125 per month on energy efficiency. In actuality, we personally are achieving $135 savings per month in our home

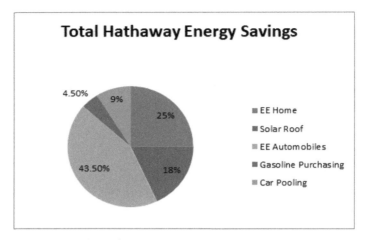

Figure 8.   Our energy savings experience.

through energy efficiency, but allowances must be made for the differences between new construction and retrofit. For example, we had a dense pack insulation put in the walls. It would be quite difficult now to add tighter wall insulation after the sheetrock is installed. With the possible exception of cold northern climates, no level of energy savings would make sense to rip off the sheetrock, put in insulation, and then replace the sheetrock and paint.

We are only $75 away from that $125 target and there are a few more things we can do to get there.

STEP 5: Increase the Home's Thermal Efficiency

Heating and air conditioning is by far your largest load making 30% to 50% of your energy consumption depending on your climate and the time of year. It's no wonder our power bills can be so much higher in the summer than the cooler spring and fall months. If air conditioning weren't such a power hog, we wouldn't see such a disparity in our bills based on the time of year. There are two ways we can focus on reducing this huge load. First, we can be a bit wiser as to how and when to heat and cool our homes. Second, when energy is expended to heat or cool something, the simplest way to increase the efficiency of that process is to ensure whatever is heated or cooled stays that way.

Smart Thermostats:

One of the ways we can more wisely heat and cool our homes is by using smart thermostats. The key to smart thermostats is their wireless connectivity and the ability to control them with your phone. The ease and flexibility of controlling your heating and cooling even when away from your home can help you add a few bucks to your monthly savings while maintaining a comfortable home for when you return.

I'll give you an example of how that works at our house. On a cold winter day, for example, I like to open all of the blinds on the Southside of the house to let natural sunlight (radiant energy) in to heat the house. I then turn down the heat or completely turn it off. With a standard thermostat, this causes problems when Carol and I return home from work after the sun has gone down on those shorter days. We likely saved 10 to 12 kWh, but the house is a little chilly. I don't mind as much, but it's never wise to cause your wife discomfort to save a dollar. With a smart thermostat, I can queue the heat to come on before we get home so it's comfortable as soon as we walk in the door. By doing this for a few months of the year when the temperatures are extreme enough to make savings worthwhile, I can save $70 per year.

So, I went shopping a couple of years ago for a smart thermostat to automate my energy-saving efforts without sacrificing comfort. I was able to learn some of the different features of different smart thermostats, many of which come with their own smartphone apps and energy dashboards. I explained to the rep I was speaking with that I already had a dashboard and didn't need to pay extra for theirs. Couldn't I just integrate the thermostat into the dashboard I already have? She looked at me like my face was upside-down.

Perhaps the technology isn't quite there yet.

But the features they do currently offer are quite helpful for energy efficiency. Some smart thermostats begin to learn your routine and automatically adjust settings after a while. Some can even detect through your phone app whether you are home or not and adjust your HVAC to save energy when you aren't home. What sense is there in keeping your home at the perfect temperature when no one is there? This is a simple and almost automatic way to save a few dollars on your monthly electric bill and will cost you $250 or less.

I upgraded our thermostats in early 2020 and was certain we were saving energy almost immediately. It turns out that this additional control over our largest load helps add an average of $10 to our monthly energy savings. But one more piece of anecdotal evidence comes from my own record keeping since 1997. It was then that we set the goal to get to energy carbon neutral. Although we have been maintaining carbon neutrality with our energy consumption since 2012 by purchasing carbon offsets, we never ever actually achieved it with our own technology until May, 2020. This was just one month after upgrading our thermostats. Many technologies were involved in becoming carbon neutral, but it was the smart thermostat that finally put us over as shown in Figure 9.

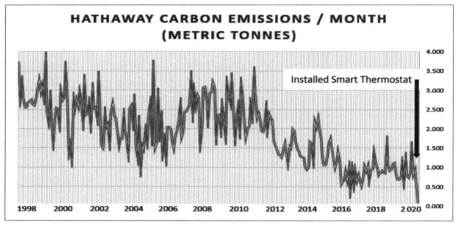

Figure 9.   Shows our ongoing progress to achieving Net Zero Carbon

Reducing Thermal Losses:

Thermal energy losses from the air inside your house moving outside can be particularly expensive, robbing as much as $20 per month. The rate that air moves from inside to outside can be tested with a blower door test, a standard requirement for new construction homes. The test uses a fan set up at one of the doors into the house to cause a positive or negative pressure to occur in the house forcing air through leaks. Pressure is monitored to determine the air changes per hour (ACH). A smoke pen will often be used to identify where exactly the leaks are for any necessary corrective action. Most state building codes require a maximum ACH rating for new homes. As a home ages, the ACH is likely to rise. You can spend around $300 for a blower door test if you would like. But we should first address common locations for air loss: doors, windows, chimney flues, and electrical boxes.

The greatest impact we can have is by installing electrical plate cover gaskets (Figure 10) and upgrading weather

Figure 10.   Gasket inserts for outlets and switches.

stripping on doors and windows where necessary. Plate cover gaskets are incredibly cheap and very easy to install. As tempting as it might be, *do not* stuff loose insulation into the electrical outlet and switch boxes. It is a serious fire hazard. Upgrading the weather stripping is slightly more of a project as you will have to ensure you have the right type for each given door or window. So, you may want to limit weather stripping to areas where damage is likely contributing to air losses. The savings you find versus the cost you pay are going to be very subjective. But, let's say all you need is an $8 pack of 20 plate cover gaskets and $50 of weather stripping and you can squeeze out another $7 in savings. That's an annual return of 145%.

Not all heating and air losses are from air moving out of the house. Your attic is likely allowing radiated heat losses, especially in the warmer months of the year. This final thermal efficiency project will require a professional. It is likely that your roof has no insulation underneath it at all. From a standard efficiency standpoint, the justification might be that your attic isn't a conditioned space. So, builders simply lay insulation above the ceiling. That's logical, but we can do better. Especially in the summer when dark asphalt shingles soak in the heat of the sun, your attic will roast. That is unconditioned space that may be home to your water heater, duckwork, or air handler! In my son's house, you can even see the tips of the roofing nails penetrating through the plywood and they are hot to the touch in the heat of a summer day. The incredible differential in temperature between your attic and the floor below is likely too much for a layer of insulation above your ceiling to mitigate. But you can hire a contractor to spray foam underneath the roof in your attic as a barrier to the radiant heat from the roof.

The ROI from spray foam is also subjective depending on the construction of your home. The cost can be $1,000–$4,000. With the median cost of $2,500 and assuming we can achieve $15 per month in savings, our annual return is 7.2%. Because of the lower ROI, though, you'll want to table this upgrade, until we get the quicker savings into our Energy Freedom Fund each month.

STEP 6: Upgrade to Energy Star Appliances

If you have standard appliances, there's an additional opportunity for savings. Energy Star (energystar.gov) is a great resource for shopping for new appliances with the intent of making a return with energy savings. If you are shopping for appliances in person, you'll notice "Energy Guide" yellow tags and stickers alerting you to the possibility of energy savings. When you are ready to upgrade an appliance, look up your current models for data on typical energy usage as your baseline before you go shopping. Then, compare your baseline with the estimated data provided on the Energy Guide tag. The monthly savings from the difference will be added to your Energy Freedom contributions each month.

I said "when you are ready to upgrade" because an appliance is more than energy savings. If you just purchased a new washer and dryer a year ago, I doubt you want to go through the trouble again, losing money on reselling the one-year-old appliances, all to chase $15 in savings per month. In our Energy Freedom investment model from Chapter 3, we assumed two appliance upgrades in year one because there will be plenty of money for appliance upgrades within six to eight months. But these upgrades can happen whenever you are next in the market for an appliance.

When calculating the ROI of your upgrade, you don't want to count the entire cost of the appliance. If we are talking about a refrigerator, for example, you are going to buy a new fridge eventually whether you want to save energy or not. Every house needs one. But you are going to spend a little extra on the fridge that performs its basic function more efficiently. You also do not want to count additional costs for non–energy-related features. If you want a cool refrigerator with cameras and a cell phone app that lets you see inside while you are grocery shopping for an extra $1,000, that's great! Don't count the $1,000 here. Only the additional cost above a basic model attributed to energy efficiency is what we figure into our ROI calculation.

Energy-efficient refrigerators have two key advantages over standard refrigerators. Predictably, they provide superior insulation so that the compressor does not have to cycle as often to maintain interior temperatures. They also often feature a freezer drawer below the refrigerator. As heat rises, cold air will fall to the bottom of your fridge, reducing the thermal losses of your freezer below. If your Energy Star refrigerator upgrade is $300 more than a similar base model appliance for $5 per month in savings, your annual return would be 20%.

Hopefully, your old top load washing machine is ready to be replaced. If so, $400 is more than enough to cover the incremental cost of going with a front loader (Figure 11). Front loaders are inherently more energy efficient. Refer to the yellow Energy User's Guide, a label posted on each appliance prepared by the EPA/DOE Energy Star Program. All new appliances should have the yellow Energy Users Guide label. This label tells you how much energy the appliance will use compared to others. This will allow you to compare for yourself and pick the unit that saves the most with the additional $400 you have to spend on such an appliance. It is less important to replace the dryer than the washing machine since the front load washing machines generally save energy both in the wash cycle and also in the drying cycle because the front load washers wring out more water. Our experience is that this change will save an additional 50 kWh per month or five to six dollars per month.

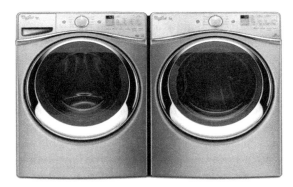

Figure 11. Front load washers and dryers save energy by removing more water from the clothes before the drying cycle.

We've already talked about wrapping your water heater, but if your water heater is ready for replacement, upgrade to a more energy-efficient one. You can get an electric heat pump water heater that replaces standard electric water heaters and reduces energy use by two thirds to three quarters saving 10 to 12 dollars per month. These can run $2,000, plus or minus a couple of hundred dollars. So, again, you are only looking at spending an additional few hundred dollars over a standard water heater. You may also decide to replace the dishwasher with an Energy Star rated one. Energy savings for a super energy-efficient dishwasher will run about 200–300 kWh per year or about $2.50 per month. And the difference from the standard to the energy-efficient dishwasher will likely be even less than with other appliances.

Heating and cooling with heat pumps is the most efficient way to condition the home. I include this in Step 6 because air conditioning and heating system sometimes called the HVAC system for Heating, Ventilating and Air Conditioning is the largest appliance (in terms of energy use) in the house and because Energy Star supports Heat Pump systems. That said we have already removed perhaps one third to half of the heating and cooling load on the HVAC system through earlier measures we have taken, such as lighting improvements, insulation, and weather sealing. The savings potential then has been cut in half. So, to make an upgrade to heat pumps work, it is necessary to focus on significant improvements in the Seasonal Energy Efficiency

Rating (SEER) for air conditioning systems or for heat pumps the Coefficient of Performance (COP).

SEER and COP ratings are usually embedded in state, city, or county building codes. For example, SEER for homes built in the period 1995–2005 was in the SEER-9 range. By 2015, SEER-13 was the standard for new home construction, but SEER-16 is now available for most systems. If you had a 20-year-old home and wanted to upgrade the A/C system to SEER-16, your A/C savings would be 43.8%. The SEER system associates incremental efficiency at each rating. So, we can easily calculate the percentage of savings by calculating the percent change from one rating to the next.

Average household energy for air conditioning is about 20–25% of the entire home energy bill or about $900 per year. So, if we are able to save 44%, total savings of about $394 annually are possible. But remember we have already saved at least one third of that from insulation and sealing and other measures, so the annual savings projection is about $260 annually. Upgrading A/C systems cost anywhere from $2,500–$4,000 making the payback at best about 10 years and closer to 15 years, so considering a full heat pump changeout might make more sense.

Heat pumps perform like air conditioners in the summer pumping heat out of the house, but in the winter a reversing valve is engaged turning the air conditioner around, so to speak, and pumping heat into the house. The benefit of a heat pump is you not only get the air conditioning savings, but you get the heating savings as well. It is harder to achieve the efficiency of a SEER-16 air conditioning rating on the heating side (4.5 COP) unless you install a geothermal heat pump. In my experience, a geothermal heat pump offers too small an ROI to be worth discussing here.

However, we can achieve between 3.0 and 3.5 COP on the heating side of the heat pump for a standard air-to-air coupled heat pump. This could reduce our heating energy costs by approximately costs by 30% to 50% depending on how well the unit and the surrounding space is insulated and sealed. Better insulated space to begin with will mean slightly smaller savings from an efficiency increase in the heat pump. Since average heating costs are around $600 annually in the south this adds another $180 to the $260 or $440 per year for both heating and cooling. The cost of a new heat pump system is in the $4,000–$6,000 range suggesting an ROI that is marginally in our target range.

Currently, the cost of adding geothermal to the heat pump adds $10,000–$15,000 more to the HVAC system upgrade price tag. Although we may now save $500–$700 per year, we are still talking about a payback of 20 years or more and an ROI well below our target. So, I only recommend geothermal heat pump systems for new construction.

After Step 5, our efforts are going to slow considerably. We are easily generating $100 in savings per month on top of our 50 dollar contribution at this point. If we have pursued all appliance upgrades discussed, we will likely have reached $150 in savings per month. But the investments remaining are going to get considerably larger. Everything we have done up to this point could be done within one year without pulling out more than 50 dollars of your own money each month. The next upgrades are going to be a few years apart from each other. Make sure savings over your baseline continue to accumulate in your Energy Freedom Fund so you can pursue the next upgrades when the time comes.

It's time to fast forward a few years.

$150

6

Drive for Free

Horses are expensive. So, by the beginning of the Industrial Revolution, the thought had occurred to mankind that travel without the need of a living animal that had to be fed, watered, and stabled was desirable. Attempts at a horseless carriage were made as early as the late 18th century – before the United States of America even existed – with steam-powered wagons. Benz, Ford, and others wouldn't be producing gasoline-powered cars for more than one hundred years! Even electric vehicles would precede gasoline vehicles, produced using lead-acid batteries in the middle 19th century. We carry this notion in the 21st century that hybrids and electric vehicles are new. However, they were ahead of their time as they beat the internal combustion engine to the market.

In the late 1890s, a fully electric car was also not an uncommon sight. But because the electric vehicle preceded the gasoline-powered vehicle, batteries were used to extend the range of gas cars making them the first hybrid vehicles. Hybrid vehicles were produced by several manufacturers throughout the teen years of the early 20th century. But the assembly line production of internal combustion engines coupled with ever-increasing power capabilities led by the Ford Motor Company by the 1920s put the early hybrid automobiles out of business.

STEP 7: Upgrade to Electric Vehicles

New battery technology which offers higher energy for less weight resurrected the hybrid electric vehicle making them much more energy-efficient than those 100 years ago. With the nickel-metal hydride battery technology powering a hybrid electric drive system, as developed for the early Prius in 1996 for Japan, and eventually brought to the US in 1999, Toyota re-launched the hybrid. Typical hybrid automobiles have about a 1 kWh drive train battery that allows for the capture of energy when coasting, braking, or cruising. For these resurgent hybrids, the primary propulsion is the gasoline engine. The hybrid battery provides energy savings and power boosts, enabling the internal combustion engine to operate more efficiently. Then, at slower speeds, you might get between three and four miles for all-electric drive on the fully charged battery before it needs to be charged again.

People that own hybrid electric sedans know they can get between 45 and 60 miles per gallon, as we regularly did with our first hybrid, a 2021 Toyota Prius (Figure 12). Hybrids in other vehicle body types also provide significant boosts in fuel economy. On average, this is a savings of up to 400 gallons per year or about $1,200. A hybrid vehicle may sell at a premium of $3,000–$4,000 over a standard non-hybrid model, which means most owners would recoup the extra price in between three and four years of driving. Typically, hybrids have MPG ratings between 35 and 50, and most efficient hybrids are about a quarter to half the emissions of the typical automobile.

Figure 12.   Our first hybrid, a 2001 Toyota Prius, parked in front of our first solar house on the National Capital Mall, Earth Day, 2001.

These hybrids are increasingly seen as the more traditional hybrid model where the internal combustion engine is the primary means of drive. In the traditional model, the only way the battery is charged is via the gas-powered engine. Even charging from coasting and braking is preceded by propulsion from the internal combustion engine. If we are going to get completely off of gasoline and drive on energy that we produce, the electric motor must be primary.

A number of car companies offer increased battery capacity with their hybrid design adding the ability to plug in and charge the battery. These are called Plug-in Hybrid Electric Vehicles (PHEVs). For every additional kWh of battery size offered on the extended battery plug-in hybrids you can expect to add an additional three to four miles of range. To gain performance and range a lighter weight, higher energy density battery technology is typically used. Lithium-ion batteries provide greater range per weight than nickel-metal hydride or lead-acid batteries. We currently own a Honda Clarity (Figure 14), which has a larger onboard battery than our first PHEV, the Chevy Volt, (Figure 13) which GM discontinued in 2019. The Clarity's range is 52 miles per charge. Because we are driving 52 miles without the internal combustion engine turning over, plug-in hybrids rate fuel economy with MPGe (e for equivalent). MPGe

Figure 13. Our first PHEV, a 2012 Chevy Volt, charging up at Northside Hospital in Sandy Springs, Georgia.

accounts for the actual gallons we might use occasionally and adds in the converted gallon equivalent of the electricity. You might think that we aren't using any gallons at all in full-electric drive so MPG should be infinite. And, it would be. But, the MPGe rating values the electricity we use to charge the battery as a sort of fuel as well to completely measure the efficiency of both forms energy put into the vehicle. The MPGe rating of our Clarity is 110 miles per gallon!

The batteries require charging either at Level 1 (120 V) for approximately 11 hours or Level 2 (240 V) for approximately four and a half hours. At 120 V the car will require about 1 kW of power for 11 hours or so. At 12 cents per kWh, the cost for a charge is $1.32 at home, although some electric charging stations can charge up to $2 per hour for charging. Some homeowners may elect to use special off-peak rates at night, if offered by their local utility, to charge their PHEVs. This could reduce the cost of electricity for charging the EV by up to 50%. Either way, there are significant energy savings using electricity over fuel.

If you can go 52 miles on a single charge for $1.32 with a Honda Clarity, what would that cost in your current vehicle? Let's compare the cost to a similar full-size sedan like the 2020 Honda Accord with a 2.0 L, 4 cylinder engine. With a combined fuel economy of 27 MPG, it would cost you $5.78 to drive the Accord 52 miles with gas at $3.00 per gallon. A 52 mile round-trip commute Monday through Friday in this scenario would save you almost a hundred dollars per month.

PHEVs qualify for a Federal Tax Credit of up to $7,500 depending on the size of the battery. Plug-in hybrids are priced at premiums of $6,000–$15,000 over conventional cars, so they can pay back faster than hybrid cars when using the full federal tax credit. Based on average experience, when a PHEV is primarily used as a commuter car, the percentage of energy used from gasoline is less than 25% while that used from electricity is 75% or more. Thus, a PHEV saves as much as $125–$175 per month compared to a conventional automobile. Paybacks can range from immediate to three years. The emissions of a typical PHEV are about 10–20% of the average automobile.

Figure 14.   Our second PHEV, a 2018 Honda Clarity.

The difference between PHEV cars and Electric Vehicles (EVs) is the complete absence of an internal combustion engine in an EV. An EV is pure electric with a much larger battery to provide ranges of up to 400 miles on a charge. Also, EVs generally can be charged at the higher Level 3 (480 volts) as with Tesla's Superchargers. Teslas can be supercharged in less than an hour. On their website, Tesla claims 15 minutes on a Supercharger will get you 200 miles. A Federal Tax Credit of $7,500 is also available, however, premiums can vary even more widely than for PHEVs. But, just about every major car company offers an EV.

The only challenge with EVs, compared to PHEVs, is the reliance on EV charging locations that are much fewer in number compared to gas stations. However, EV manufacturers mitigate this as charging infrastructure continues to spread and mature. Many of the luxury EVs, as well as higher-end Teslas, provide an onboard navigation system that provides trip planning including charging station locations along the way. EV drivers also become very familiar with EV charging station locator apps such as Plug-Share to find their next charging point. An EV will save on average $200–$250 per month in energy needs. After tax credit, they pay for themselves almost immediately or at least within three years.

After everything we have talked about so far in this chapter you may be wondering how you can save $200 or more with one EV. You are probably doing your own math, too. An average car at 25 miles per gallon driving an average 14,000 per year only costs $140 to fill up per month at three dollar a gallon. So, where does that other savings come from? I have found that with even one EV in our garage, we have not only replaced one internal combustion engine, but more than one. With few exceptions, when driving together, Carol and I opt for the EV every time. So, our EV completely replaced the miles I used to drive in a gas-power car, but it also replaces some of the miles driven in the remaining gas-powered car.

Other EV Opportunities

Promoters of new electric vehicles are pointing to the ability of the EV to be a generator for power into the home. Ford markets its new F-150 electric truck as a backup generator, for the home when the power goes out. But imagine if you were on a real-time rate with the electric utility. You would now have the option with your EV truck to power the home when electric rates were high, then recharge the truck when rates were low. This could be a game-changer!

From Gas to Electricity

We spend less than $20 per month for electricity to charge our plug-in hybrid to drive an average of almost 1100 miles per month. This is an incredible value, at less than 1.9 cents per mile. No one has traveled on less than two cents per mile since gasoline sold for $0.50 per gallon! Not only are EVs quick paybacks, but they offer extraordinary ways to boost your monthly savings making it possible to accumulate significant cash quickly for the next investment: a solar energy system.

One of the greatest advantages of owning an EV (including plug-in hybrids) is the fuel can literally be free when we generate our own electricity with a solar energy system. Even better, the EV will actually help make the pay back on the solar even faster. This is because utilities often do not credit energy exported to the grid at the true market value of solar energy. After installing a solar system, if there is a month where we consume less than we produce, depending on the utility, we will receive a wholesale credit (more on why that is a problem in Chapter 8). This can be as little as one third of the value we get when we consume energy from our solar system. However, an EV provides the opportunity to soak up any excess solar energy and put it to use in our vehicles, thereby preserving the value of the solar energy output. It allows you to store the excess solar energy in your EV's battery for use when you need it. This extra value to your PV system can increase the rate of return on the solar investment by one and a half percent or more. For example, we are currently getting a 10.75% ROI on our solar system, up from 9.25%, by having an EV.

Our Journey Away from Gasoline

I have always looked for the best gas prices before filling up. On the typical road trip to Maine every summer with the family when our kids were growing up, I would plan out which states we would stop in and which states we would drive through, if possible. This is pretty typical. I'm sure you would drive a mile past a gas station close to empty if you knew you could save a couple of dollars at the next one. But I have always taken saving money on gas further than most. I began aggressively hunting for savings on gas in 2004 using websites and apps (like Gas Buddy) that track gas prices at every pump. As prices began to increase and grow more volatile in the Fall of 2005 following Hurricane Katrina (see Figure 15)[6] we stepped up our gasoline purchasing plan.

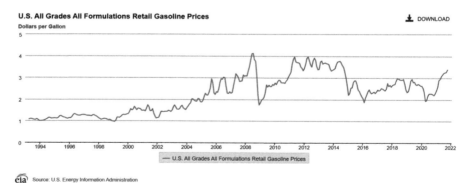

Figure 15.   Historic gas prices over 30 years. Volatility in prices increases following the end of 2005.

---

[6] https://www.eia.gov/dnav/pet/hist/LeafHandler.ashx?n=pet&s=emm_epm0_pte_nus_dpg&f=m

I would search out the lowest cost gasoline in the area and go purchase our gasoline for the week. It became my weekend ritual. Just by doing this, I was able to save 10–15% off the average price in the area, living in Virginia at the time. You may even consider hunting down the best prices in your area to inject a little more savings at the beginning into your Energy Freedom Fund.

As our gasoline consumption dwindled with our first hybrid and now our plug-ins, I still look for the best way to buy gas for as little as possible. But, by driving a plug-in, I can spend even less on a gallon of gas than I ever could before. When you lower household gasoline consumption by 50% or more due to the use of PHEVs and EVs it makes it easier to purchase total monthly gasoline at special prices with fuel programs at grocery stores. We like to use the Kroger Fuel Points Program. Because we fill up so infrequently, but our spending on groceries hasn't changed, we can amass enough points in a month to get $1 per gallon off of up to 35 gallons of gasoline. We signed up for the double points program with Kroger that enables us to get to that $1 off per gallon with $400 in grocery purchases each month plus four store surveys. Sometimes, I reach $1 off twice in one month. So, I can get up to $1 off per gallon for 70 gallons each month. On average, though, we only need 50 gallons, a month.

Someday we won't need any gas at all.

But, if we just talk about the gasoline we buy, not even counting the miles we drive with no gasoline consumption, our savings are still substantial. The chart below shows the average national price for gasoline as reported by the U.S. Energy Information Administration. Over the past 20 years, the price has zigzagged between $1 per gallon and $4 per gallon. Adding the 10–15% savings by price shopping, we started using grocery fuel programs after moving to Georgia in 2007 before realizing in 2014 that Kroger had the best one. Taking the average price per gallon in Virginia and Georgia, I have compared our own average price per gallon with the savings we have achieved. As a result, the average price per gallon we have paid throughout our nearly 20 years of gasoline purchases is about $2 per gallon. And we have saved $5,600 just in how well we purchase gasoline. We have enjoyed half of that savings in just the last six years as the gap in our actual paid price vs the average price in the region has widened as can be seen in Figure 16.

Driving an EV, even a plug-in hybrid EV, has been more than just about savings. EVs liberate us from the chains of gasoline, especially when we create our own fuel at home. Just a few months ago, as of the time of this writing, the Colonial Pipeline cyberattack affected 50% of the supply of gasoline in the Southeast. Just a couple of days after the attack, gas stations

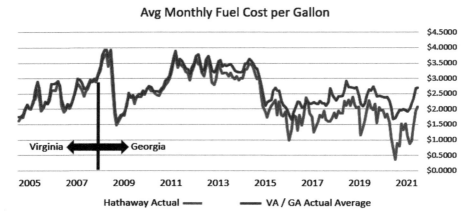

Figure 16.   Shows our actual paid price per gallon trend vs the Average for Gasoline in Virginia and Georgia respectively. The gap has widened as we made greater use of grocery fuel reward programs.

began running out of gas. I counted four out of every five of gas stations on my 45-minute commute home that dried up. Driving my fully charged PHEV, I drove by those stations with a smile as we conserved the gasoline we had at home for Carol's car.

7

Own Your Energy

We are about to make a critical turn.

Up to this point, we have been enhancing the efficiency of our energy consumption. This has generated savings which have contributed to more upgrades, begetting yet more savings. We now maximize the usefulness of energy while minimizing our consumption. Boosted in performance, the locomotive races full speed down the line needing little fuel to maintain momentum. But it still needs fuel.

Now it is time to shift from energy efficiency to energy generation. And, as of the writing of this book, there is no better technology for near-universal household energy generation than solar photovoltaic (PV) energy.

Solar PV and How it Works

Solar photovoltaic literally means "sun light electrical." Solar PV systems are systems with the components necessary to convert sunlight into usable electric energy. These systems require only a few key components in order to generate electricity for your home: PV modules (solar panels, as most people know them), inverters, electrical enclosures (panel boards, disconnects, etc.), and wire and conduit.

PV modules, the most recognizable part of the system, are typically made of many identical crystalline-silicon cells that are laminated into a rectangular module, stamped with conductive material, and framed for structural mounting. When sunlight hits the silicon, electrons in the silicon are excited and captured in the conductive material that you might say is painted on the cells. The conductive material connects the cells to each other and brings the electricity to a junction box on the module. Out of the junction box comes a positive and negative wire. Modules are mounted together in what is called an array and the module leads are either connected to one another or directly into micro inverters.

Solar modules naturally produce DC power. But because our homes are built around an AC power infrastructure, we must change the DC power into AC power through an inverter. Inverters come in different sizes and configurations, but their role is the same. In one way or another, the leads from the PV modules must get DC power to an inverter (or inverters). Then we can connect the inverter's AC output to our AC electrical panel to power the home.

There are many different types of components in PV systems. For example, instead of the traditional framed silicon module, there are thin-film PV products that don't use silicon and can be made in the form of roofing shingles or flexible adhesive strips, however, these are uncommon and often cost-prohibitive. Whatever technology is used, sunlight must be converted into DC power, and DC power must be inverted into AC power.

Why Solar?

You never hear of anyone putting a nuclear reactor on their roof, or a coal-fired steam turbine in their back yard. But, in 2020, despite COVID-19, 3.2 gigawatts (GW) of solar PV systems were installed on the rooftops of homes in the US. That is more total power output than the tenth-largest power station in the country, the Turkey Point Nuclear Generating Station in Miami-Dade County. And, 2020 wasn't a special year for residential solar: it has been growing exponentially for decades now. There are two key reasons solar PV is gaining so much steam as a growing source of power, and the only feasible source of power that almost anyone can generate at home.

First, PV systems are modular and easily scaled with great flexibility of design. A solar PV system can be tailored to the site conditions and the energy and power requirements of the facility or home it will power. PV modules can be added or subtracted and come in different sizes. Likewise, modules can be connected in different ways into various sizes and types of inverters before tying into the existing electrical panel. A single solar PV system can be large enough to rival the output of centralized power plants or small enough to fit on your roof and power your home.

Secondly, solar power is renewable. There are no mines, drilling rigs, pipelines, or tanks to refill necessary for a PV system to generate power. It just has to receive the free gift of sunlight. So long as the sun is in the sky, we will have power. This also makes the economics of solar energy more attractive. Once the up-front cost of the system is paid, the fuel is free and the maintenance is minimal. Systems are expected by the industry to last 30 years or more. Amortizing the up-front cost over the system's useful life, solar PV pays for itself many times over. But, your LED lights, efficient appliances, and electric vehicle already paid for your PV system.

What about diesel, gasoline, propane, or natural gas generators? These generators serve only as a backup for when the power goes out. They do not make sense for daily use, because the fuel cost and ongoing maintenance costs would reverse all the good progress you've made. With fossil fuel based generation, you would be retreating from Energy Freedom back into bondage - this time to a supply of fuel that you can't control. A solar system runs daily with no fuel costs and never needs an oil change or parts repair. To take it a step further, with the addition of a battery energy storage system, a solar PV system can even replace a fossil-fuel-fired generator as a backup means of power.

Solar Myths

Solar energy, while not a new technology, is near the center point of a shift in the energy industry. However there are a lot of myths out there that we still need to address. Here are the five most common as they relate to residential solar PV:

Myth 1: "PV systems do not work when it is cloudy. FALSE." Cloudy weather does block some of the sun's light, but there is usually plenty of light for the system to produce power on a cloudy day. Depending on the weather and the clouds, the total energy produced on a cloudy day will be lower than on a sunny day, as should be expected. Other weather effects impact production as well. Interestingly enough, cold weather actually improves production. And, if there is snow on the ground, the reflection of the sun off of the snow added to the direct sunlight will boost production. When sizing your PV system for your energy demand, designers will take into account these weather-related variances based on average solar irradiance for your area.

Myth 2: "PV systems can power my home when the power goes out." FALSE. Alone, PV systems cannot produce power when grid power is lost. This is because standard solar inverters match grid voltage and turn off when the grid voltage is not present. It's a good thing, too, because utility workers nearby might not know if a PV inverter is energizing the line when attempting to repair that circuit in the outage. But with the addition of energy storage and a few other components, we can turn our PV system into a microgrid that will isolate itself from the grid and continue to generate power when the grid goes down.

Myth 3: "Solar energy is too expensive." FALSE. The ability to produce your own energy is an incredible thing and should cost something. But as briefly mentioned already, the upfront cost of producing your own energy is well worth the decades of free energy. There's even more to this myth surrounding the value of the energy you get out of your investment, but we will need another chapter to address it.

Myth 4: "If I have a PV system, I won't have a power bill." FALSE. Even if you never purchased another kWh from the utility, you would still be charged the basic fees for having an account and being connected to the grid. But chances are you will buy energy from the utility just as you will sell energy back when you produce more than you are using. You can almost be guaranteed that there will be a bill of some sort. But, stay connected: your participation in the electric grid as an energy producer is a key part of our new energy economy.

Myth 5: "My system will become obsolete as technology improves." FALSE. The solar industry is growing fast which begets technological innovation. But a system installed today will not suddenly stop working when a new module comes out with higher efficiency, for example. Waiting for technology to "arrive" will have you waiting for eternity all while missing out on the savings of producing your own power now.

Myth 6: "Solar modules are not very efficient." FALSE. This myth likely refers to module efficiency which is usually around 20%. But that doesn't matter. All this means, roughly speaking, is of the photons traveling at the speed of light that hit the module, 1 in 5 will force an electron out of its atomic orbit and onto the conductor. We don't care about photons; we care about electrons. So, module efficiency, which matters to manufacturers seeking to create better technology, doesn't matter to us. We will still design a system that gets us the energy we need.

Myth 7: "Solar may work down South but not for me in the North where there is less sun." FALSE. It is true that there is generally more solar irradiation in southern locations versus northern locations, but solar irradiation is only one factor in the economics of investment in a solar PV system. If you combine other factors, such as the cost of electricity from the grid, incentives, system design, etc., one could find myriad examples where a northern solar array outperforms a southern one on the basis of a return on investment."

"There is another element at work here. We have already pointed out that your Energy Freedom efforts act like a financial investment portfolio. When all of these improvements are done in concert with one another, the ROI of the overall portfolio improves. You will also find that the difference of a complete Energy Freedom portfolio in ME versus one in GA, for example, has less to do with the sunlight that falls on one of the roof."

STEP 8: Install a Solar PV System

All of the work that we have done so far culminates at this turning point. By now we have built an energy savings engine that generates around $350 per month in energy cost reductions. So, if we are still making that $50 per month contribution, we are putting away nearly $5,000 every year. Not only have we built an energy savings engine that will help us take the pivotal step of owning our own generation, but we have eliminated waste in our consumption that no longer requires additional generation to feed our bloated demand. Some of the folks that claim myth 3 above have likely paid too much to generate energy that is wasted.

This is an important point that warrants emphasis. The ability to generate your own energy is an awesome thing. Awesome things come with a cost. Why, then, do so many people pay more of that cost for a larger PV system to power their 60 W light bulbs when it could have saved them thousands of dollars to spend a few hundred on an LED upgrade? If you were offered a thousand dollars in exchange for a hundred, wouldn't you take it? Of course, and that is why we have spent time upfront to make sure our locomotive is at cruising speed and running as efficiently as possible.

This step can be as easy as getting three quotes from local solar installers and hiring one to install your system. Be sure to find a company with NABCEP certified installers that will install your system. NABCEP is the North American Board of Certified Energy Practitioners and they hold the standard for solar PV professionals. You should also ask your salesman if they are PV technical sales certified.

However, it will behoove you to know to a small extent technically what they will be proposing.

The most important part of quoting a solar PV system is knowing how big it needs to be. In order to do that, the companies quoting for you will start with your consumption. Expect them to ask for 12 months of your utility bills. For our purposes, we will assume a home in Georgia uses 15,000 kWh annually. 1,250 kWh in average monthly consumption[7]. But, remember we have reduced our bill by at least 750 kWh per month in our energy efficiency work (Chapters 3–5). Then, we added back in about 300 kWh per month for charging our EV at home (Chapter 6). It ends up being about 800 kWh per month or about 27 kWh per day. We care about daily consumption because the sun rises daily and we will use data on the sun's daily solar irradiance compared to our daily use to determine how much solar we need.

---

[7] https://www.eia.gov/consumption/residential/reports/2009/state_briefs/pdf/GA.pdf

We can find solar irradiance data (Figure 17) at the www.nrel.gov,[8] the website for the National Renewable Energy Laboratory. Atlanta, GA gets an average of 4.75 kWh per square meter per day of sunlight potential.

Sunlight potential shown on the irradiance map takes into account weather that impacts the sunlight that will reach our solar array. We can tell by the fact that the zones on the map do not cut straight across latitudinal lines. Southern Arizona and Georgia are roughly at the same latitude and the sun is in the sky the same number of hours per day. So, shouldn't irradiance be the same? Taking weather data into account, the answer is no.

Knowing the irradiance in Atlanta, GA, and our daily consumption, we can begin to figure our solar PV system size. If you remember from solar myth 6, we can expect a module efficiency of about 20%. The exact efficiency will be determined by the module the salesman includes in his or her proposed design. The software he or she uses will automatically take this into account. But at 20% efficiency, we will get 20% of the potential energy that the sun irradiates or 0.95 kWh per square meter per day. By dividing the daily consumption of 27 kWh by the daily production of 0.95 kWh per square meter we get 28.5 square meters or 306 square feet. You might think this is the required area for our solar array. But we need to make one key adjustment.

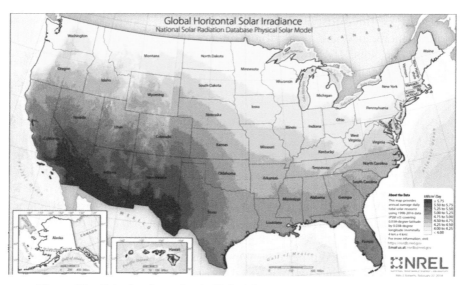

Figure 17.   Solar irradiance for the United States varies by latitude and climate.

[8] https://www.nrel.gov/gis/assets/images/solar-annual-ghi-2018-usa-scale-01.jpg

Because we use AC power, our daily consumption is an AC figure. The solar array will generate DC power before being inverted into usable AC power. Because there are some small losses from the modules to the electrical panel, we will need to increase the size of our array a bit to get the 27 kWh per day that we need. Losses can come from high temperatures, slight variances in the output of modules in the same string, inverter efficiency, long wiring runs, soiling of the modules, and other factors. The proposals you receive should take these factors into account. Conservatively, let's assume these losses add up to 20% for easy math. If we lose 20% of what we produce on each square foot of our array, we should add to our array size. By dividing 306 square feet by 80%, we come up with an array size of 383 square feet.

Referring to the solar irradiance map, you'll notice that the same energy output will require more space in Boston, MA, and less space in El Paso, TX.

There are other design factors your salesman may incorporate into the proposal that they should be able to advise you on. These may include the following:

Clipping: Clipping is a design technique that over-sizes the array and/or under-sizes the AC inverter within manufacturer specified limits to widen the production curve. As the sun comes up more and more energy is directed at the array until noon when the energy begins to reduce again. If a larger array feeds a smaller inverter, the inverter can reach its maximum power sooner and hold it longer. This is not required but is common in the industry for overall system performance.

Tilt: For most residential systems, the tilt of the array will be determined by the slope of the roof. Tilt is important because the sun shines at particular angles in the sky throughout the year. If the average angle of the sun in the sky is 34 degrees, an array tilted at 34 degrees will produce the most. For ground-mounted systems, other factors like wind load may mean that the ideal tilt for production isn't so ideal overall.

Azimuth: This is the direction that the array faces. Again, your roof may determine this for you. If your roof faces slightly to the West or the East, so will your array. Maximum production can be achieved at an azimuth of 180 degrees, or due South. However, other generally south-facing arrays will perform well, too. If the orientation of your roof has you deciding between an Eastward or Westward azimuth, consider the time of day when your load is the highest. Then choose the

azimuth that orients the array to where the sun will be during that time
of day.
String Sizing: Strings are a group of modules connected together in parallel
as one single input into the inverter (does not apply to micro inverters).
Somewhat depending on the physical limitation of the array area, string
sizing can impact the voltage input in the inverter which affects the
energy generated by the system.

As of the time of this writing, it would be common to find modules that
produce 18 watts per square foot. Given our array area of 383 square feet,
we would need a system with a nameplate rating of approximately 7 kWDC
(kilowatts of direct current). What 7 kWDC "nameplate" means is that the
output listed on the nameplate, or sticker, on all of the modules in the array
add up to 7 kWDC. Whether or not there is a module on the market that
divides perfectly into 7,000 watts or if the number of them can be connected
in the right strings for whatever inverters are on the market is a problem for
the designer to solve.

Ask your salesman questions about how these factors have influenced
the proposed design. The intent is not to worry so much about how we are
getting the 27 kWh per day that we need. Rather, we want to know that we
can trust that the company making the proposal is thinking through these
factors properly.

Your twelve months of power bills will include the power that flowed
to your EV's battery. But, as an exercise, how much of your monthly energy
consumption is your EV responsible for? It depends on several factors. The
size of the EV battery, its kWh capacity, is the primary factor and closely
related to the vehicle's range. However, the battery may need more energy
to recharge if it has been fully discharged. The amount of daily driving you
do also affects the energy your vehicle draws from your home. If you drive
a short commute and never fully discharge the battery, you won't need as
much energy for charging. And, if you are able to charge away from home
(at work, at the grocery store, etc.), that is energy that won't appear on your
power bill.

For me, I can charge at work, so I only need to charge enough for a one-
way, 25 mile commute. One kWh is enough energy for the average EV sedan
to go 3.5 miles, so I need just over 7 kWh per day. Adding in other driving
we do, we average 9 kWh per day for charging. For us, this means about
2 kWDC of our solar PV system supports our electric driving.

Solar PV Cost and ROI

As of 2020, the cost of residential solar in the US averaged about $2.50 per watt installed. At that cost, the pre-incentive price of a 7 kWDC system would be $17,500.

Don't get sticker shock!

There are many factors that influence what your return on investment might be: system design, your physical location, your state, your utility, your rate schedule (how you are billed by your utility), your tax liability, just to name a few. Many of these factors relate to utility rebates, special rates for valuing your production, and state and federal tax incentives. We will not discuss the specifics of these because they are highly variable and may even change before you finish reading this chapter. Just know that the companies competing for your business will make sure to show you all of the various ways you can recover some of the cost of your installation before pulling out your checkbook.

However, to give you an idea, as of the writing of this book, there is a 26% federal investment tax credit (ITC) for those that install a solar PV system. In this case, the credit would take $4,550 off of the new solar owner's tax bill for the tax year. Now the out-of-pocket cost has dropped to $12,950. At nearly $5,000 per year in contributions, you have the cash for the system in two years and eight months.

Most investments in solar PV are met with more incentives than just the federal ITC. For example, your state may also have a state income tax incentive. Your utility may also offer a rebate or a special rate schedule for valuing the energy produced by your system. Why? Utilities want to attract solar on their grids for different reasons. Some see the benefit of distributed generation and its implications for the future of the grid. Others may be required by their state's public service commission to have a certain percentage of renewables on their grid. Either way, the utilities can help buy down the up-front cost of solar with rebates or increase the value of generation through a special rate schedule. Most utilities offer net metering for solar systems. With net metering they allow the solar system to generate the negative energy credit against the positive energy use through the meter throughout the 30-day billing cycle. At the end of the month, the net of usage is what is billed for. Sometimes there is a negative balance because the solar system has generated more energy during the month than the electric consumption at the house. In this case, utilities settle up by purchasing the excess solar with a credit on the bill, usually at wholesale electricity prices, $0.035 per kWh. But some utilities will carry the negative balance forward to the next month which provides for full net metering throughout the year.

Let's just assume that a 26% ITC is all that there is. Let's also assume that we can get the full "net-metered" retail value of energy that we export to the grid (for example, when we produce more than we use – think mid-day when it is sunny, but everyone is at work and or school).

According to the Energy Information Administration, the average retail electric rate for residential customers into 2022 is 13.82 cents per kWh (although we have been using $0.10 per kWh throughout this book – so we will stay conservative and continue using $0.10 per kWh). If you are the average customer, this is the rate you would pay the utility for each of your 27 kWh every day. But, if our solar PV system generates a kWh, that is one less kWh we pay for. Generating exactly what we need, an average of 27 kWh per day, our PV system generates $2.70 per day in savings or $985.50 per year.

Your system pays for itself in 13 years. However, over the course of those 13 years, electric rates will rise. Rates have a strong history of escalation slightly above inflation at the rate of 2.7% per year since 2001. Over those 13 years, they will average at least $0.115 per kWh so savings will actually average $1,100 per year. System payback will be in the 11.5–12-year range with an estimated ROI of 8.5%.

Again, this assumes no incentives beyond the federal ITC and the full value of your solar energy, which isn't asking much. It also conservatively assumes that the value of your produced energy only increased 2% per year. If it still sounds like you are waiting a long time to recover a lot of money, remember: all of your efforts up to this point have already paid the cost. If you never started this journey you would be spending $350 more right now for your energy and your Energy Freedom Account wouldn't have $12,950 in it. But you DID start.

All of those efforts have added up over the few years since. It's as if the You from Chapter 3, from the mere decision to start, jumped in a time machine to now (most likely in eight years or less if you followed the plan precisely) and handed you $12,950 for your PV system.

Don't stop now!

Figure 18. Our first solar home in Hillsboro, Virginia.

Figure 19. Our second solar home in Cumming, Georgia.

$500

8

What is the Value of Energy?

In Chapter 7, we tabled much of what can be said about the myth that solar is too expensive. The truth is that this myth is bigger than solar. It is based on the faulty assumptions we make when valuing energy. In fact, the answer to this myth reveals that this key flaw feeds the status quo, so our energy future depends on setting it right.

Most utilities today treat all kilowatt-hours – the unit of electrical energy – the same. We know from Chapters 1 and 2 that monopoly utilities prefer to serve load with as much large, centralized generation as possible. This means that fixed costs for the enormous power plants make up most of the cost structure. A little bit of extra fuel and staff to pump out more energy doesn't make much of a difference to the cost to produce that next kWh. So, when negotiating power purchase agreements with qualifying facilities in accordance with PURPA, utilities compare the rate they will pay the generator with the cost of producing energy in their own large plants where marginal cost is fairly stable and likely very low. For a qualifying facility to earn a power purchase agreement from the utility, it must provide energy at the same rate or better on average over its lifetime. It doesn't matter what type of generator it is or how it will serve load. A kWh is a kWh. This view of the value of energy is called the levelized cost of energy (LCOE).

LCOE ignores the opportunities of different types of generators to meet demand in different ways. This is a critical failure of LCOE because demand is not constant. Think about when we demand power. Our lowest demand is at night, of course, when we are sleeping. Depending on the time of year and the climate, the demand profile can change, but demand is higher during the day. When does a solar PV system generate energy?

You already know the answer.

Critics of alternative energy sources, particularly solar, argue that they are too expensive. Their position ought to surprise no one because the cost of any new source of generation on the grid is compared to incumbent utility generators according to its LCOE and not according to the market value of the energy produced.

Consider the "Duck Curve." Figure 1 comes from the California Integrated System Operator (CAISO), an organization that helps manage and coordinate the grid across multiple utility territories. The Duck Curve, appearing in the shape of a duck, shows CAISO's projection of a load profile that is getting more and more volatile (Figure 20). You can see that energy is in higher demand between the hours of 6–9 am and again from 6–9 pm. In perfect competition, this would affect the price of energy because generators would have to dispatch more of their capacity to meet the demand. In turn, the higher price at peak times would reduce the quantity demanded, flattening the curve and maintaining equilibrium. Conversely, prices would drop in hours of low demand as generators look to keep selling energy and avoid shutting down.

There is a bit of irony here. The CA Duck Curve is caused primarily by too much solar energy generating all at the same time. Monopoly utilities pursue alternative energy and distributed generation mostly out of mandate or PR pressure as there is no internal incentive. So, this chart provides a convenient opportunity for utilities to say, "See? Solar doesn't work!" But the truth is quite different.

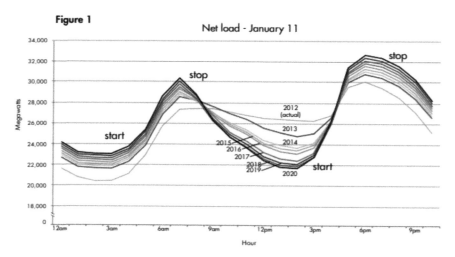

Figure 20.   The California Duck Curve – Net Load on the Electric Grid as reported by California Independent System Operator (CAISO).

The cost or value of solar energy, isn't the problem. The model that California utilities have used to promote solar clearly doesn't work. And California has only led the way. This problem will begin appearing in states across the country where utilities have also incentivized the addition of solar on their grids without the proper price signals. Almost always, the LCOE model for valuing solar generation is to blame. For utility-scale solar, the LCOE hurdle is the marginal cost of incumbent utility generation, or the wholesale price (called "avoided cost" in the industry). Beat that price, the utility will sign a deal. For distributed generation on homes and businesses, in most cases, LCOE is based on the retail value of energy that the customer avoids paying the utility or the retail price. Either way, a kWh is a kWh and it doesn't matter to the utility, or the customer installing solar, when it is produced.

Asking solar to compete with the utility's internal marginal cost is like asking a featherweight to box in a heavyweight match. It's the wrong match for a featherweight to box in, but monopoly utilities will let the fight go on and then convince you that the featherweight is the inferior boxer. This makes absolutely no sense.

So, rather than designing solar projects that attempt to generate solar energy when the energy will be most valuable, solar project developers focus instead on maximizing production. If a kWh is a kWh, there is no incentive to give up a small amount of production in order to point the array SW for production slightly later in the day when demand is a bit higher. There would also be no reason to add an expensive energy storage system that can hold onto solar energy until the grid needs it most.

The sun will shine when it will shine. Solar is not a perfect match for most load profiles which typically peak in the later afternoon. Solar production peaks at noon assuming an array facing due south. So, energy storage compliments solar very well with its ability to hold onto energy produced and discharge it at the optimal time for the grid and for the system owner. In my first solar home that we built in 2001, I had a bank of batteries next to our inverter in the basement. When the power went out due to Hurricane Isabelle, our lights were on because the batteries stored all of the energy generated during the day so we could use it when we needed it. Batteries are not advanced technology. They have been around long before the Duck Curve.

Energy storage technology is even being used in California, but not to its fullest extent because the price signals telling when it is most valuable don't exist. A recent editorial by Gerald Braun, a former colleague of mine, writing for the Integrated Renewable Energy Systems Network (IRESN)[9], said;

"Are California's energy resilience assets being used to provide energy security for diverse and important groups of individual electricity customers?

[9] https://www.iresn.org/news/2021/7/29/resilient-decarbonization-requires-state-and-local-leadership

Yes, but not for most residential and commercial customers. Has deployment of on-site energy resilience assets in California over many decades enabled numerous energy resilient communities? Not yet. Are energy resilience assets being integrated with grid assets to maximize local energy security? Not yet. They are not called on to feed electricity into the regional grid when it is under stress. Two strategic opportunities are being missed. First, the opportunity to use local resilience assets to back up the state's electricity system when California power plants and imports from other states fall short of meeting aggregated demand. Second, the opportunity to isolate and continue to serve local areas cut off from the state's electricity system due to regional or localized blackouts."

In short, California already has the means to fix the Duck Curve except for the ability to mobilize them effectively whereby balancing load and generation. I asked Gerry if the California utilities were likely to offer a means of real-time net metering. He said that a utility rate system based on rewarding users to supply energy back to the electric grid equitably remains unlikely in California. Thus, where it is needed most, there won't be real-time pricing anytime soon.

STEP 9: Achieve Energy Freedom with Energy Storage

You may have noticed that the Energy Freedom investment model from Chapter 3 does not include energy storage. That is because an energy storage system (batteries, grid-forming inverter, controls) does not provide energy savings back to us the same way all of the other energy upgrades to our home have. Despite that, energy storage is the lynchpin in Energy Freedom. You can never operate autonomously without it. A backup generator will power your house when the grid is down, but you'll be right back to grid dependency after a couple of days of burning fuel.

Yet, energy storage hands you complete control over all of your energy. Thus, its value is not savings, but true freedom via autonomy and resilience as your home becomes a microgrid. The definition of the term "microgrid" generally refers to a larger system of loads, generation, and storage that can "island" from the grid; it is not usually meant to describe a self-sufficient home. But you have all the same components and islanding capability, so I think the term applies.

An energy storage system today can cost $12,000–$18,000, but is eligible for the same federal tax credit as your solar PV system when they are connected together. At about the 10 to 11 year mark after starting toward Energy Freedom, your fund will likely have what you need to complete this final step. We have also experienced rapidly dropping costs in batteries in the last few years as rising demand for EVs has meant more batteries must be produced. This helps the energy storage industry build economies of scale. So, chances are very good that by the time you reach this point, your energy storage system will cost less.

Because storage does not provide energy cost savings the same way as all of the other upgrades we have discussed, it doesn't make sense to calculate an ROI as we have been doing. Instead, it's best to understand the different ways energy storage provides additional value. Beyond the autonomy and resiliency, you will have, storage may do even more:

No Net Metering:

If you do not have access to a net metering program with your utility, energy storage can help maximize the value of the solar energy you produce. As discussed in Chapter 7, net metering provides your solar system generation with full retail value as a statement credit. If you use a kWh of your solar energy, you avoid paying the utility the retail rate you would have paid without solar. But, if you send a kWh to the grid, you accept what the utility is willing to pay for the energy. Without net metering, the utility will only pay wholesale or avoided cost: likely one third of the retail rate or about $0.035 per kWh. As always, this depends on your utility.

It is not uncommon for your solar energy to go onto the grid. Remember, your solar PV system produced all of its energy during the day, and it peaks at mid-day. There is likely not enough load when everyone is at work and school to use all of that energy. So, it goes to the grid during the day and you buy it back from the grid at night. You can assume that as much as 60% of your solar energy is mismatched with your load, meaning 60% is exported to the electric grid. Think about the hours of day you use energy the most in your home compared to the hours that your solar PV system will produce energy. That energy must go somewhere. With net metering, the utility acts as your battery because the value of the energy you sold is the same as the value you will buy later. If net metering is not available, your energy storage system can hold onto your solar energy until you are ready to use it later in the day.

In terms of dollar savings, the value of being able to achieve full retail rate net metering with battery storage is worth approximately 40% of our anticipated solar value, or in the case of our solar array from Chapter 7, $37.50 per month, $450 per year. This value is not additional to the solar savings we calculated in Chapter 7, it is the amount that batteries will preserve in the savings if net metering is not available.

Energy Bill Arbitrage:

It is common for utilities to also offer alternative residential rate schedules as well as the fixed-rate most of us are used to. If there is a rate schedule with time-of-use rates or on- and off-peak rates, then you should change your service to those rate schedules. The key is that you want rates that are higher during one part of the day and lower in another. That way, you can hang on to energy produced by your solar PV system that might otherwise go to the grid when the cost to buy from the grid is low. When the rate is higher, you can use the free energy from your batteries effectively increasing the value of your savings from the grid.

Peak rates are sometimes valued at four to five times their off-peak rate ($0.25 per kWh vs $0.05 per kWh), so it can increase the value of your savings by using batteries to minimize when (hopefully to zero) purchasing electricity. It makes sense to use your own generated electricty when rates are their highest and only purchasing any needed electricity when rates are at their lowest. My calculations suggest that we can increase the cost-saving of our solar system by 10% by switching to real-time rates with a battery system – or about $10 per month.

In a way, when you have energy storage, you get to decide the value of your energy. Energy nirvana would be achieved here if the utility offered real-time net metering so that you could receive the full peak value of electricity by exporting your stored up solar energy to the electric grid at $0.25 per kWh.

Demand Response

When the load spikes, utilities have to find ways to serve it by firing up auxiliary generation. Or, they can incentivize customers to reduce their load. Some demand response programs operate on the premise of customer load reduction. However, tapping into energy storage as a means of serving load during high-demand times is an increasingly popular demand response tactic. Owners of energy storage systems can enter into contracts directly with the utility to provide their stored energy when needed the most. However, these programs designed specifically for energy storage are yet uncommon.

For residential customers, the closest we have to this demand response program is the air conditioner cut-off programs during the summer. Utilities will incentivize you to cut off your air conditioner during the summer, and this is when you could deploy your battery/microgrid to kick in and drive your air conditioning system. This credit is worth $24 to us for each of the three summer months of the program.

If we tallied these values up for energy storage, we end up with an additional $640 annually.

The ROI, then, is about 4.3%. This is well below our target of 7%, but then we never calculated the value of electric resiliency – being able to have power when the electric grid goes down.

Even with the value we can harvest today through energy storage today, the future smiles on the technology. Our grid recognizes the need for distributed energy storage and will find new ways to open the energy market to it. The frontier is here and just over the horizon lies a new energy economy.

As I hinted in Chapter 6, there is a new EV Truck technology being promoted by Ford Motor Company, the F-150 Lightning, which is being advertised as a battery for the house on wheels. Who needs a stationary battery when I can purchase a truck that has a battery capable of operating our solar home for four days or more. In fact, I have ordered the F150 Lightning to round out our complete energy system and get us one step closer to that new energy economy. When the truck will be available is anyone's guess, but not before May. Stay tuned!

9

The "Why Not" Mindset

"Tuvalu is sinking."

That's what the people of Tuvalu lament as their small island nation slowly goes underwater, according to a 2019 article.[10] Many of us are comfortably unaware that many people in Tuvalu may be forced to evacuate permanently from their homes. The 11,000 citizens of Tuvalu may eventually become ecological refugees seeking a new homeland as a result of global climate change. Appealing to other nations for a new home, they had been turned down by Australia as of the time of this writing. However, New Zealand has started granting resident status to Tuvalu families on climate change humanitarian grounds beginning in 2014, and families have begun to migrate to higher ground.

I have been aware since the late 1980s of the claims about global warming due to the so-called "greenhouse effect" caused by the release of carbon dioxide into the atmosphere at ever-increasing rates, trapping more and more solar radiation. I was also aware of the links between sulfur dioxide and nitrogen oxide emissions to acid rain and smog. Back then, I used the link to tout energy-efficient lighting as a way of helping to reduce our environmental impact. There was a lot of opposition (even within my own company at the time – Sylvania Lighting) to using the environment to promote energy-efficient lighting. Apparently, our relationship with the environment encroaches on some serious ideological territory. My attitude back then was more or less indifferent to the environment.

"Whatever your attitude about the environment, it should not obscure the fact that we can sell more energy-efficient lighting to the environmentalists as well as the environmentally-minded businesses. Furthermore, we can say that a company's investment in energy-efficient lighting not only pays back with an attractive return on investment but, it's good for the environment, too."

---

[10] https://www.theguardian.com/global-development/2019/may/16/one-day-disappear-tuvalu-sinking-islands-rising-seas-climate-change

I would say Why Not to my Sylvania colleagues simply trying to meet out another selling point for our products. Many companies I worked with adopted this Why Not strategy as well. We weren't trying to save the world; we were trying to save people money. The environment was just a cherry on top. I call this the "Why Not" mindset. Why not pursue energy savings with the mutual intention of reducing our carbon impact?

This interlude about climate change may seem sudden. Aren't we just talking about Energy Freedom for our household? Likewise, in my early lighting days, I was just trying to sell new lighting technology that would help our customers save money. But there is a larger benefit to what you are doing, even if you are as indifferent as I was about climate change.

However, by the mid-1990s, I began to notice that there was something very real about the claims of global warming, the more simplistic term we used back then for the phenomenon. Many today still doubt that our climate is changing. This is astonishing to me as I am personally a witness to it. The island in Maine where I have vacationed since 1959 has an old piling of stones that were used for a dock long ago. When I was a young boy, we used to play by those pilings, walking out to them at low tide. At the highest tide, the stones were never covered. Today, there is not a high tide that doesn't completely submerge them. Even since the 1990s my son has noticed a clear change in water level in his shorter time visiting the island. So, I further investigated the projections and the historical facts about increasing carbon dioxide concentrations in the atmosphere, carbon emissions from human activity, increasing temperatures, and rising sea levels from melting arctic ice. I arrived at the conclusion that human activity is clearly altering the global climate.

In addition to researching the issue on my own, my personal experience validates what I've found. When we lived in the Baltimore-Washington metropolitan area from 1993 to 2007, it was ranked in the top four worst areas for ground-level ozone or smog. Although traffic congestion and point sources of emissions of nitrogen oxides were growing and causing a significant amount of ozone, the Maryland Department of Environment claimed as much as 50% of the ozone came from the combustion of coal for electricity. In the summer of 2002, 2400 people were admitted to local emergency rooms complaining of asthma .on an unusually high number of severe ozone alert days for the region.[11] Over 200 of those people admitted died. However, from

---

[11] https://www.washingtonpost.com/archive/local/2003/01/23/running-out-of-breathing-room/aefa3f2d-f027-4e49-ae2f-f80bcf086a09/

2000 to 2012 in the Appalachians, just to the west, we saw a boom in wind development that gradually replaced a lot of that old coal combustion. By the summer of 2013, the region's air quality had improved dramatically,[12] proving that renewable energy can make an important difference.

So what? Ground-level ozone is not climate change and climate change is not ground-level ozone.

When ground-level ozone is formed from the burning of fossil fuels, so are greenhouse gasses (GHG). GHG, like carbon dioxide, are directly linked to climate change as their presence in the atmosphere influence how much solar radiation is maintained inside the atmosphere instead of reflected. More carbon dioxide means more heat is retained. The carbon atom is the predominant atom in fossil fuels as fuels like coal and oil are made from very old organic (meaning carbon-based) material. When carbon is burned, a chemical reaction combines the oxygen in the air (necessary for combustion) with the carbon atoms to form carbon dioxide.

Ozone is a separate by-product of burning fossil fuels. Ozone is a compound of three oxygen atoms, versus oxygen, which is a compound of two oxygen atoms. It is not a direct by-product of fossil fuel combustion. Almost all fossil fuels contain small amounts of nitrogen. Similar to the burning of carbon, nitrogen is also bonded to oxygen in the air to form different combinations of nitrogen and oxygen that are emitted from smokestacks and tailpipes. As these molecules blow around in the air, heat and light from the sun break them up such that oxygen atoms released from their bond with the nitrogen form bonds with other oxygen atoms to form ozone.

Here's why this matters. Carbon is, by far, the dominant atom in fossil fuels, so GHG emissions are much greater than ozone emissions. The fact that ozone was causing hospitalizations in the DC area in the summer of 2002 attests to the incredible amount of GHG that had also been released. Imagine the reduction in GHG emissions when even the ozone noticeably retreated as wind power displaced coal. If you appreciate the reduction in GHG emissions you will inevitably cause, then you already have the "Why Not" mindset. Why not pursue Energy Freedom with the mutual intention of reducing your carbon impact?

By this point, you have already dramatically minimized your carbon footprint. There is nothing more you need to do except possibly quantify exactly how much your carbon footprint has shrunk.

---

[12] https://www.washingtonpost.com/news/capital-weather-gang/wp/2013/09/26/breathing-easier-washington-d-c-s-remarkable-improvement-in-air-quality/

Your Carbon Baseline

In the same way we calculated our energy cost baseline at the beginning, each of us should calculate our carbon baseline. What was our carbon footprint when we first started to move toward Energy Freedom? In their book, "Our Ecological Footprint," authors Mathis Wackernagel and William Rees explain that the average annual per capita American emission of carbon dioxide is 19.5 metric tons, nearly 43,000 pounds. According to The Consumer's Guide to Effective Environmental Choices, written by Drs. Michael Brower and Warren Leon for the Union of Concerned Scientists, energy for household consumption and transportation is about 67% of total GHG emissions per household.[4] This is calculated based on our average energy consumption, including the gas we combust in the engines of our vehicles and the emissions that come from fossil fuel burning generation to deliver that energy. The 33% of our GHG emissions not directly associated with our energy consumption occurs because of the food we eat and the manufacture of the products we consume. Also, according to Drs. Brower and Leon, the 67% emissions due to transportation and home energy are almost equally divided: 32% and 35% respectively. Again, this is average. We can figure our carbon impact according to our lifestyle.

First, it is important to understand the types of carbon impacts we cause. In 2002, the World Resources Institute convened a panel of experts at the famous Rockefeller estate on the Hudson River in White Plains, New York to help define the Greenhouse Gas (GHG) Protocol. I was honored to be among some of the panelists. We determined that carbon impacts are broken out into three different scopes:

Scope 1: Direct Carbon Impacts
Scope 2: Indirect Carbon Impacts
Scope 3: Supplier Network Carbon Impacts

Direct carbon impacts occur when you own the tailpipe, chimney, flue, smokestack, or other devices from which carbon is released. Examples of Scope 1 impacts are, your car, your fireplace, your natural gas or oil furnace. In other words, you and I are the only parties causing the emissions of our scope 1 carbon. Indirect carbon, on the other hand, is emitted by your energy supplier. It's not your smokestack or mine, but we are using the energy created in the process. The cause of emissions of scope 2 carbon is shared.

Supplier network carbon impacts are a bit of a catch-all. All other carbon emissions fall in this scope. This carbon is emitted by third parties on our behalf to produce products we buy: food, clothing, toys, appliances, etc.

This includes the cost of getting raw materials and final goods from their point of origination to our doorstep or to store shelves. Interestingly, these are scope 1 and scope 2 emissions for the companies that produce these products or that transport them. As such we will not estimate our scope 3 carbon impact.

Calculating our carbon footprint can be really simple, especially since I will give you the emissions per unit in each scope.

Scope 1[13]
Gasoline: 18.74 pounds of carbon dioxide per gallon.
Natural Gas: 12.06 pounds of carbon dioxide per therm.

Scope 2
Electricity (Depends on Region): Use an average 1 pound per kWh

Once you have a total number of pounds for a month, divide the number by 2204.6 to convert to metric tonnes. Why metric tonnes? The international community has settled on the metric tonne of carbon as the universal unit for measuring carbon dioxide emissions. A metric tonne is slightly more than 20% larger than a US ton.

We can still factor in an assumed scope 3 impact. Calculating it would be impossible. But we know 33% of our impact is not from our energy consumption. That is the same as saying 33% of our carbon footprint is from scope 3 impact. All we need to do is take our total scope 1 and 2 impact and divide it by 0.67 to get our total impact. The difference between our total and just scopes 1 and 2 is our scope 3 impact.

Knowing this baseline, you can easily track how your carbon footprint is dwindling as your savings are rising. Tracking the reduction of your carbon footprint may be ancillary to your pursuit of savings and energy autonomy. But, it's easy to do and I encourage doing this in order to know the full impact of your Energy Freedom.

Carbon Offsets for Carbon Neutrality

If like me, you recognize the threat of climate change and have a desire to eliminate your personal contribution to carbon in the atmosphere, there's another easy step we can take. The ultimate goal for mitigating carbon is complete carbon neutrality – a carbon footprint of zero or negative pounds of carbon emitted from your activity from scopes one and two. As we have

---

[13] https://www.eia.gov/environment/emissions/co2_vol_mass.php

already said, the highest reasonable expectation for our energy reduction is 80% from our baseline. That still leaves 20% of the time where we will still burn a little gas, or buy a few kilowatt-hours from the grid. We can offset this residual carbon impact by purchasing inexpensive credits that offset that impact we can't feasibly eliminate on our own.

We are not seeking a return on investment here as with any of the energy upgrades we made. Achieving carbon neutrality for the individuals and organizations that buy them is an expense made as a matter of living up to their vision of promoting renewable energy or fighting climate change. It's a choice we make because of what we care about. A veteran-owned company Tripp previously worked for offers a relatable example of this motive. As a veteran-owned company, they wanted to attract and retain veteran employees. Doing so did not require that the company give a portion of its profits to non-profit organizations that cared for veterans. But, they did because that's who they were as a company.

Some purchases of carbon offsets are a bit more compulsory, however. In some states, regulated utilities have a Renewable Portfolio Standard (RPS) mandated by the state requiring a certain percentage of their generation to be renewable. If utilities with an RPS don't have enough true generation, they can purchase Renewable Energy Certificates (RECs) to cover the balance and meet the requirement. This is not to say a utility that purchases RECs has little regard for their carbon impact. They very likely do. It simply means that a requirement has been imposed that is designed to increase renewable generation, mitigate carbon, etc. Similarly, private companies seeking funding may need to consider their carbon impact and develop a plan to reduce it. Some large investment firms require this for access to funding because they increasingly view sustainability as a characteristic of good, investment-worthy business.

If we are going to offset our scope 1 and scope 2 carbon emission, you probably guessed that there are two categories of offsets.

Scope 1: Carbon Dioxide Verified Emission Reductions (VERs or Carbon Offsets)

Scope 2: Renewable Electricity Certificates

These credits and offsets don't exist just to make us feel good. They actually do something, whether reducing carbon, increasing renewable deployment, or both. Carbon Offsets, negating our scope 1 emissions, are measured and verified reductions of carbon dioxide resulting from a project that was created with at least one main purpose for reducing carbon emissions to the atmosphere. Carbon offsets must be projects that occur outside of usual business activities and be verified by a third party and registered in a carbon registry or tracking system. Carbon offsets are considered the gold standard for carbon accounting, for they can be used to offset any scope 1, 2, or 3 carbon impact as long as you can boil that impact down to tonnes of carbon dioxide. Since the basic unit of a carbon emission is the metric tonne, the unit of a carbon offset is one metric tonne. Typical prices are seven to twenty-five dollars per tonne. The purchase of offsets helps finance these projects that actually reduce carbon in the atmosphere.

Because we are not directly responsible for the carbon emitted to produce the electricity we purchased from the grid, the carbon offset is not the best purchase for mitigating scope 2 emissions. For scope 1, you can easily calculate your impact; you know how much gas you put in your tank. For scope 2, you are probably not going to call the utility and ask them which power plant your kWh came from last month and how many tonnes of carbon were emitted to produce them. But your power bill tells you precisely how many kWhs you used. RECs represent electricity produced by a renewable energy generator (solar, wind, biomass, etc.). The unit of production is one MWh per REC, or 1000 kWh. RECs typically cost half as much as carbon offsets. And, again, when purchasing a REC, you are helping to finance renewable energy projects, attracting more renewable generation.

Our GHG Impact Since 1998

In our household in 1997, we had five individuals living under one roof and operated two automobiles. Due to my job and our distance from relatives our family traveled much more than the average family. Between flights and car travel, we logged almost 120,000 miles in 1998. The average emissions from energy consumption expected for a family our size would suggest something in the neighborhood of 72 tonnes of GHG emissions per year or six tonnes per month. However, our own energy-based GHG environmental impact back in 1998, when we began this journey, peaked at more than nine tonnes of carbon dioxide per month, but averaged about 7.2 tonnes per month. So, our 1998 household baseline was about 87 tonnes of carbon dioxide emissions per year.

Just as we calculated added savings from our baseline that financially fuel our energy upgrades, we also should calculate our reduced carbon emissions from those savings. For us, transportation made up an above-average share of our overall carbon footprint. So, as we upgraded to more fuel-efficient vehicles, to hybrids, and EVs, our carbon footprint shrank noticeably. And, since we know that our emissions are directly tied to the number of gallons of gas we burned, if we reduce our gasoline consumption by increasing the fuel efficiency of our cars (Figure 21), of course, we reduce our carbon footprint. We already have achieved the savings and the flexibility to drive on gasoline or electricity as the market and supply dictate. But now, we know that we have reduced our carbon dioxide emissions from over two tonnes per month (225 gallons) to 0.59 tonnes (66 gallons), a 71% drop.

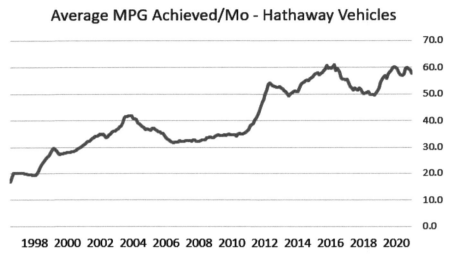

Average MPG Achieved/Mo - Hathaway Vehicles

Figure 21.   The increased fuel efficiency of our automobiles since 1998, has resulted in a 71% drop in fuel consumption.

Our solar house is achieving even more spectacular results. Our current monthly energy demand from the grid is averaging about 97 kWh in 2021, down from 1871 kWh per month in 1998. That's incredible by itself, but when we consider that average includes 271 kWh for charging the PHEV, then the house is performing as a net negative electricity home. Only our natural gas at 29 therms per month has increased since 1998. Even so, our average emissions from consuming natural gas and electricity combined are a mere 0.196 tonnes per month. This compares to approximately 1.2 tonnes per month in 1998 – an 84% reduction.

Altogether, we have reduced our scope 1 and scope 2 GHG emissions from about 45 tonnes per year to 9.4 tonnes per year, down nearly 79% from mid-1998 (Figure 22). Even when considering our total GHG emissions, including emissions associated with food and the manufacture of the products we consume, we still have reduced our total GHG impact by over 50%.

All of this has been accomplished while improving household energy performance and raking in savings.

Why not?

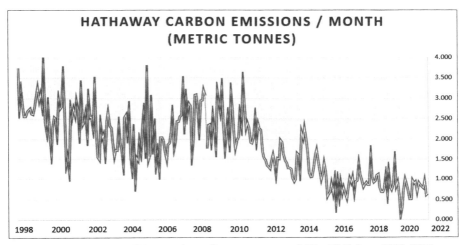

Figure 22. Our GHG impact from all energy sources fell by 79% from 1998–2021.

10

The New Energy Economy

Congratulations! You have accomplished each step of the road to Energy Freedom. Your locomotive is cruising down the line on its own power and you are beholden to no one. No grid outage will shut you down, foreign and domestic policy on energy won't affect your wallet, and you are likely saving $450 or more per month (Figure 23). Yet, there is even more good to come of your journey. Because you produce and manage your own energy, you have the ability to become a participant of the electric grid. When you started this book, you were merely a customer or ratepayer. As of the writing of this book, the extent to which you can actually participate in the grid as a generator is limited. So, we are entering a bit of uncharted territory.

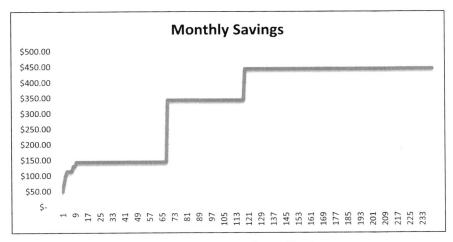

Figure 23.   The steps to Energy Freedom.

From here, things can get very interesting!

Someday, you will trade energy with your neighbors at prices you both agree on. What would that look like? Only imagination limits the possible. But I have a picture in my head. Let's say you are on vacation for a week for

Thanksgiving. You have set your heat to come on at 62 degrees and set the water heater on a strict schedule. You have also programmed your microgrid controller to arbitrage the cost of energy. This means that the batteries will keep the free solar energy and buy from the grid when the price is low, and release the energy to the grid when it is higher, earning a credit for energy supplied. You happen to see in your utility app that the price of energy is spiking on a particular day while you are out of town. Maybe a lot of homes in your neighborhood are hosting guests and using their entire homes to entertain for the holiday. The higher demand from your neighbors means a higher price they will pay for electricity.

What do you do?

A 2016 White House Report[14] indicated that there was the potential to capture as much as 17% behavioral energy savings by providing real-time feedback on the impact on utility costs. The premise was that electricity consumers simply need to see their utility costs at the time of decision. So, when you get the real-time notification that the price of energy is going up, you simply open your home energy management app and shut off the heat and water heater entirely to sell more energy to the grid at a higher price. You can turn them back on when prices normalize so that the house is comfortable by the time you return.

This is what the new energy economy will probably look like. And it won't just be neighbors buying and selling from one another; whole cities, including commercial and industrial facilities, can be connected to this modern grid and make real-time decisions about their consumption and generation to maximize their benefit. Energy users on this future grid also will not be required to generate. They can simply be users and manage their consumption in the most cost-effective way. However, as a generator, you will be a vital part of the new energy economy because you will be able to sell energy to the grid at the market price. Add the revenue from energy sales to the grid to your $450 per month in savings!

In no way does this future energy economy replace the need for a utility. Even transmission and base load generation will still be necessary to a degree, at least in the early future. But the grid will be leaner with only the minimum centralized generation to efficiently serve the base load, and even that may completely disappear someday. We are already building more decentralized generation stations like natural gas co-generation plants, solar

---

[14] www.whitehouse.gov/sites/default/files/microsites/ostp/submetering_of_building_energy_and_water_usage.pdf

fields, and wind farms. Even nuclear is getting smaller with nuclear micro-reactors just now hitting the market.

As utilities give up some or all of their generation to third-party user-generators like you, they will take on new responsibility as the broker of third-party transactions. Technology will enable a real-time understanding of load and generation capacity to set prices according to what is being demanded and what can be dispatched. Smart metering, cloud computing, and grid operation will converge in ways that haven't yet been conceived to create a free energy market that operates in real-time. Every generation meter and every consumption meter will report via 5G (6G?) to a system that can interpret the exact market conditions for energy on any given distribution circuit.

So far, this is only a dream of the future. But the dream is not far off. We already have some of the technology and programs needed to make this dream reality. Since it will take you several years to achieve Energy Freedom, it is likely that by the time you reach that point more advancement will be made in the direction of this new energy economy. For now, let's talk about the technology that we do have that can either be used in the new energy economy or as stepping stones to get there. Be on the lookout for these developments.

The Smart Energy Home

Once an idea out of science fiction, smart homes are here now. But smart *energy* homes are still developing. We discussed earlier how a smart thermostat could help you save energy with the ability to program your heating and air conditioning with your cell phone. But there are so many different ways we can control things in our homes with smart technology, that aren't necessarily energy-related. We can control our lights, answer the front door, arm an alarm system, open the garage, play music, turn the volume up on the TV, and more all from a smartphone or even just our voice. Soon, we may even be able to program our EVs in the garage to provide power to the house or even export to the electric grid. Opening the pod bay doors can still be a challenge, however.

You may have heard the term the "internet of things" or IoT. It simply describes the trend that everything you can think of will someday be connected to the internet for the purpose of exchanging data and signals that provide us with better control. This can be as simple as a power outlet with an integrated chip that connects to the cloud where a cell phone app with a simple on/off button merely replaces the need for a physical switch on the wall. Or it can be as complex as a security system that integrates multiple video feeds with data storage and connection to an emergency call center. And when I say "everything you can think of" I mean just that. Maybe there's a bit of a fad with IoT as a novel trend when we have products like the smart salt shaker invented to measure out the perfect amount of salt and provide mood lighting at the table! The point is that everything is increasingly connected to the cloud which gives us greater understanding and control. When it comes to energy, this is a critical trend.

Imagine what we could do if all of our energy consumption, production, and storage were integrated into one energy management app? We could automate certain functions based on certain conditions. For example, all lights could be automatically turned off at 8 am on weekdays after everyone has left for the day. When the batteries are fully charged, the electric vehicle charging station will be allowed to start. Or, if we are planning a party with lots of people, we could schedule the air to come down a couple of degrees just before guests arrive.

The implications for IoT as it relates to energy management in our home tie directly to our description of the future where we will buy and sell energy in the new energy economy. When the market price for energy changes, all we will need to do is pick up our phones or even ask some smart device on the kitchen counter to make some adjustments to our consumption and energy storage settings to optimize our energy performance. Real-time market pricing for energy will complement our capability for real-time or even programmed energy management.

Smart Meters and Energy Dashboards

In the Energy Policy Act of 2005, electric utilities were encouraged to begin installing smart meters to be ready for the potential of real-time electric rate pricing and two-way feed monitoring to account for onsite renewable energy generation, and have the ability to support energy diagnostic information for critical loads. Smart meters are the building block of the new energy economy and are critical to its fruition. However, much of what we need smart meters to do can already be done. They are even in service in nearly all utility service territories on loads and generators.

One of the key benefits of the utility move to smart meters has been the proliferation of software-based systems that report data that smart meters collect. It's not enough for smart meters to collect data. We need that data organized into meaningful information to foster an understanding of our consumption and production. Energy monitoring platforms and dashboards serve this function. Energy dashboards today can pull in data from smart meters and other sources as well to generate a more holistic energy picture. For example, we can get data from our solar and battery inverters detailing our production and we can also get information on critical loads like air conditioning and EV charging.

The availability of energy dashboards, with their ability to receive and display energy information from smart electric meters, points to something that seems like the logical next step: control. At least for residential energy users, this has yet to evolve. The level of control that coalesces into a single dashboard has not yet been adopted by utilities nor demanded from home-owners. However, we can see the use of energy management systems used for major commercial and industrial loads where dashboard systems allow for direct control of critical loads. Scaling these down into simpler residential energy management systems is only a small next step.

Because there is no single dashboard for homeowners, even just for data, I use a couple of disparate dashboards to keep me informed of my performance. PowerEnfo (Figure 24) by Empower provides us with immediate energy consumption feedback so that I can associate savings with a specific action. The simple "Energy Used Today" dial quickly informs me of our total usage at a specific time. Unfortunately, it compares our home to itself instead of average homes near me. Understanding our performance in relation to others will bolster our interaction with others in the new energy economy. This concept has already been pioneered, but cannot amount to much without a real-time and free energy market.

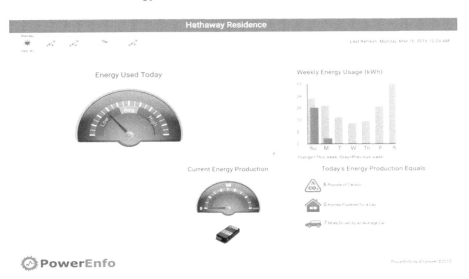

Figure 24. PowerEnfo Dashboard by Empower provides immediate information on relative energy use and generation from our solar.

I also use a dashboard that tracks my solar PV production in real-time and compares it to my consumption. In the first screenshot below, you can see my energy production from the solar (green) and my energy consumption

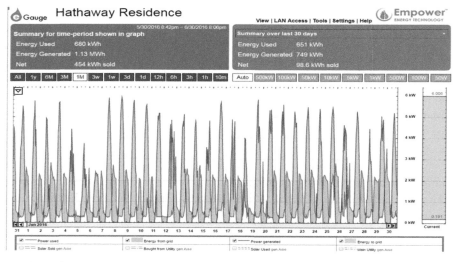

Figure 25.   EGauge Dashboard by Empower shows actual solar production in Green against energy consumption in Red over a period of 1 month (June).

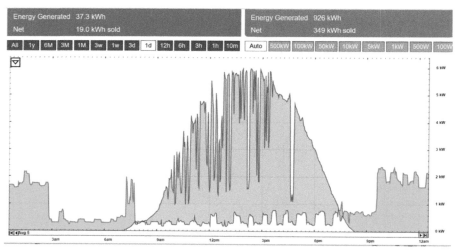

Figure 26.   EGauge Dashboard shows Solar production and Energy Consumption for single day in June.

(red) as depicted in Figure 25. This view of an entire month of June, 2016 shows that we have produced far more than we have consumed. The following view (Figure 26) shows an actual day in August 2021. You'll notice that the energy production mostly occurred when we were away from the house indicating a mismatch between energy supply and demand on the electric lines. This illustrates the need for energy storage to align solar production with demand on the grid and prevent contributing to volatile load on the grid such as California's Duck Curve.

Utility Programs

Even today with many regulated utilities, there are hints of the new energy economy that is coming. Programs designed to help customers save on their power bills have been around for decades. Newer programs look to achieve that same end but with a modern angle. Several utilities have recently begun providing free smart home devices for their customers. These free devices typically include smart outlets, smart light bulbs, smart thermostats, and even entry and motion sensors (possibly for awareness of open doors which has more to do with energy savings than home security). Free devices for free energy savings – great! Why would your utility do this?

Their answer – and a true answer – is to keep their customers as happy with their service as possible. Even as monopolies, it is in regulated utilities' best interest to do so. But there is more to it than that. Depending on the utility, the climate, the time of year, etc., spikes in load may be associated with times when people are at home. For example, if you remember the Duck Curve from the last chapter, two factors that exacerbate the typical spike in daily load in California are the drop-off of solar production and the increase in energy usage in homes as people return from work and school. These smart home programs affect household energy use, which would ease late afternoon and evening peaks such as is characteristic in the Duck Curve. Utilities know that a flat load profile is cheaper and easier to serve than a volatile one.

A similar program that has been popular in recent years gives utilities access to customers' air conditioning. Customers receive a credit in exchange for the utilities having the option of shutting off the customer's AC unit for a short period of time alleviating a spike in load. This was a precursor to demand response, mentioned in Chapter 8. Like a demand response, the intent is clear. Utilities need to balance load and generation while customers are usually willing to trade a credit for the unlikely event they will lose cooling to their home for 15 minutes or half an hour.

So far, these programs are mostly attempts to manage load or even control it to a degree. I was recently in a position to ask a number of these utilities about whether they were taking their positive experience from many of these programs and applying them to an interactive program with dynamic rate pricing, even incorporating solar and battery storage. They responded that this was a great idea, but might be some time before they put it in place. I am encouraged by this response because it confirms two things: this new energy economy is coming and utilities have an interest in it as well.

Until then...

This new energy economy will solve even the most persistent market problems that today's grid cannot solve with the status quo. The perfect competition in generation and real-time pricing that will come with the new energy economy will make the most efficient use of generation while making the load more manageable. That means no more failed nuclear deals that cost half a state billions of dollars. No more blackouts that stretch 400 miles and across international borders. No more miscues between energy pricing and project development for the new generation that fails to serve load when needed most.

Imagine what this new energy economy could have done for New Orleans after Hurricane Ida! Even if there isn't enough generation capacity on the local circuit to power the city as usual, a real-time, market-based utility could signal to the generators that *are* available that energy is at a premium and desperately needed. Maybe the price increases eightfold to $1 per kWh in an extreme event. This would allow the market to triage its need for energy. At the extremely high prices, many people may still choose to light their homes with candles freeing up the precious bit of generation capacity to be used in a hospital to save a life. Some call it price gouging today when gas prices respond to natural disasters in similar ways, but it is really the most equitable way for a market to function. The extreme prices in major events command supply for those that need it most.

You will help make all of this happen. In fact, this new free energy market is not possible *without* you. Not only have you achieved Energy Freedom for your household but you are positioned to usher in the new energy economy and even profit from it. No one knows exactly how it will unfold. So, keep an eye on the utility market and be ready.

Until then, enjoy your Energy Freedom!

Epilogue

It is quite possible for any household following this plan to save as much as 80% off of its baseline energy costs, as we have been able to achieve. At least for now, savings beyond 80% is highly unlikely, requiring you to be off-grid and make some radical compromises in your transportation. I am not recommending that. So, I consider an 80% reduction in baseline to be a great accomplishment. And, we didn't quite achieve that with our first solar home because we experimented a bit with energy design. Our second solar home provided the opportunity to make improvements. EV technology was still developing for much of our journey as well, so we did not achieve the larger transportation savings until later. But this plan provides you with all our lessons learned and fine-tuned techniques upfront.

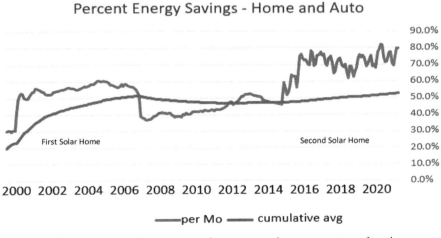

Figure 27. Our cumulative energy savings expressed as a percentage of total energy purchased 2000–2021.

The chart above (Figure 27) tells the story of our Energy Freedom journey, focusing on the last 20 years, although I spent previous years on

101

efficiency. We moved into our first solar home in Virginia and bought a hybrid automobile in 2001. When we moved to Georgia in 2007, we had to sell our Virginia solar home and live in a modest house while we watched the housing market melt down and slowly recover. By 2012, we had bought our first PHEV, the Chevy Volt. You can even see a bump in our savings with the addition of the Volt. By 2016, we had moved into our second solar home. Notice how much more we are saving in the second solar home with a PHEV than the first with just hybrid automobiles!

Our peak energy cost savings in the first home was only 60% in 2004. Our peak energy cost savings now has edged just above 80%. Also, notice how variable our energy savings are. We are operating with such narrow tolerance that even a short trip to South Carolina to see family vastly changes our total savings in a given month where we might purchase a few gallons of gas. Yet, despite these occasional trips we stay well above our previous performance in the first solar home.

By 2016, we have seen an average energy cost savings of between $600 and $700 per month (Figure 28). We are living comfortably in our large house with two automobiles and able to go where we want, without an impact on our lives from energy supply shortages or price shocks. All things considered, it took us 18 years to achieve Energy Freedom. Could it have been done quicker? The chart below shows says "yes." After we left Virginia in 2007, we sold our first solar house and essentially started over. But, within eight years we were able to return to a solar house and this time with a PHEV. We built on the lessons of the past and achieved nearly an 80% reduction in

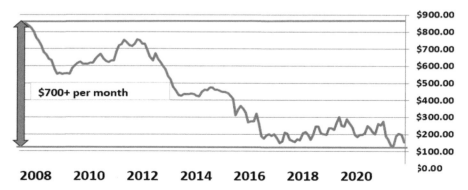

Figure 28.   Our average monthly savings upon moving to Georgia and fine-tuning what we learned in Virginia with our first solar house and early hybrid automobiles.

overall energy costs. That timeframe is more likely what you will experience. I have already completed the task of figuring out what works best and in what order improvements should be made. All you need to do is execute.

What do we do now that we have achieved Energy Freedom? Since we used to spend $1 per hour on energy, we now have $0.80 per hour with which to do whatever we please: maybe invest into retirement. Carol and I still have several years before retirement. But, when we achieved Energy Freedom five years ago, let's assume we had 13 years to invest our $600 every month in energy savings. Earning the historical S&P 500 rate of 10.5% annually, that $600 per month would leave us an additional $180,000 in retirement.

Financial advisors always make the case for young people to start investing early. It is no different with Energy Freedom. I cannot stress this enough. The extra $180,000 will be a nice bump for our retirement. But, if our children, who are in their thirties, started their eight-year journey to Energy Freedom today and invested the savings until they retire, each one of them could have $1.6 million in retirement from their energy savings alone as depicted in Figure 29.

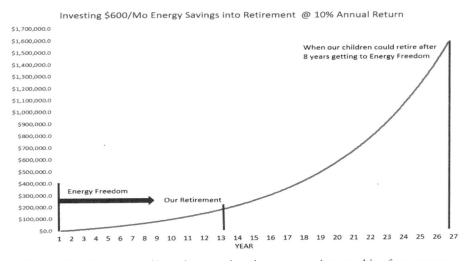

Figure 29. The power of investing over time the energy savings resulting from energy freedom.

If you know a young family, a new homeowner, or a college graduate, share this book with them. If you wish you had read this book (or wish I had written it) when you were younger, I'll leave you with an old Chinese proverb. "The best time to plant a tree was twenty years ago. The next best time is today."

Index

About the Authors

Alden Hathaway, II, supports solar development and energy efficiency programs for national accounts and utilities as Senior Vice President for Sterling Planet. He works with customers to integrate energy technologies to help them achieve carbon neutral goals while maximizing savings and return on investment.

Mr. Hathaway is an Electrical Engineering graduate of the University of Virginia and assisted the USEPA in the launch of Green Lights, Energy Star Buildings and the Green Power Partnership. Mr. Hathaway also served on a White House task force on Climate Change in 1993 and led the First Lady's Village Power mission program to provide solar powered light to East Africa in 2000. In 2001, Mr. Hathaway's net zero solar house was featured on the National Mall and in the President's National Energy Policy.

Mr. Hathaway co-founded Solar Light for Africa in 1997 with his father and the Cranberry Isles Community Solar Association in 2020, and serves on both boards. He is a registered Professional Engineer in Georgia, where he resides with his wife, Carol, in their second net zero solar house.

Tripp Hathaway, has worked with his father to achieve Energy Freedom since he was in the seventh grade. A husband and a father of his own three boys, he is now working on his own household.

As a small business executive in the energy industry Tripp worked for over a decade building solar energy systems and microgrids for the US military and large industrial energy users. In that role he defined "energy security" for the company's clients as the value they would receive by investing in microgrids. To Tripp, energy security could only be obtained if an energy project included sustainability, energy resilience, and cost savings. Tripp believes that, to a homeowner, Energy Freedom closely resembles energy security.

In addition to Tripp's work in energy, he serves as an Air Defense Artillery officer in the Army National Guard. He has been on three tours of duty with service in Kosovo, Germany, Poland, and Turkey.